Public Perception of Tap Water Chlorinous Flavor

The mission of the Awwa Research Foundation (AwwaRF) is to advance the science of water to improve the quality of life. Funded primarily through annual subscription payments from over 1,000 utilities, consulting firms, and manufacturers in North America and abroad, AwwaRF sponsors research on all aspects of drinking water, including supply and resources, treatment, monitoring and analysis, distribution, management, and health effects.

From its headquarters in Denver, Colorado, the AwwaRF staff directs and supports the efforts of over 700 volunteers, who are the heart of the research program. These volunteers, serving on various boards and committees, use their expertise to select and monitor research studies to benefit the entire drinking water community.

Research findings are disseminated through a number of technology transfer activities, including research reports, conferences, videotape summaries, and periodicals.

Public Perception of Tap Water Chlorinous Flavor

Prepared by:
Erin D. Mackey, **Hélène Baribeau**, **Ana Cristina Fonseca**, **Jason Davis**, **Jess Brown**, **Lina Boulos**, and **Gil F. Crozes**
Carollo Engineers
12592 West Explorer Drive Suite 200, Boise, ID 83713

Philippe Piriou, **J.M. Rodrigues**, **M. Fouret**, and **Auguste Bruchet**
Suez Environnement-CIRSEE
38 rue du Président Wilson, 78230 Le Pecq - France

and

David J. Hiltebrand
AH Environmental Consultants, Inc.
Director of Water Programs, 804 Omni Boulevard, Suite 201, Newport News, VA 23606

Sponsored by:
Awwa Research Foundation
6666 West Quincy Avenue, Denver, CO 80235-3098

Published by:

Awwa Research Foundation
Advancing the Science of Water®

DISCLAIMER

This study was funded by the Awwa Research Foundation (AwwaRF). AwwaRF assumes no responsibility for the content of the research study reported in this publication or for the opinions or statements of fact expressed in the report. The mention of trade names for commercial products does not represent or imply the approval or endorsement of AwwaRF. This report is presented solely for informational purposes.

CONTENTS

* On CD Rom

TABLES

FIGURES

FOREWORD

The Awwa Research Foundation is a nonprofit corporation that is dedicated to the implementation of a research effort to help utilities respond to regulatory requirements and traditional high-priority concerns of the industry. The research agenda is developed through a process of consultation with subscribers and drinking water professionals. Under the umbrella of a Strategic Research Plan, the Research Advisory Council prioritizes the suggested projects based upon current and future needs, applicability, and past work; the recommendations are forwarded to the Board of Trustees for final selection. The foundation also sponsors research projects through the unsolicited proposal process; the Collaborative Research, Research Applications, and Tailored Collaboration programs; and various joint research efforts with organizations such as the U.S. Environmental Protection Agency, the U.S. Bureau of Reclamation, and the Association of California Water Agencies.

This publication is a result of one of these sponsored studies, and it is hoped that its findings will be applied in communities throughout the world. The following report serves not only as a means of communicating the results of the water industry's centralizes research program but also as a tool to enlist the further support of the nonmember utilities and individuals.

Projects are managed closely from their inception to the final report by the foundation's staff and large cadre of volunteers who willingly contribute their time and expertise. The foundation serves a planning and management function and awards contracts to other institutions such as water utilities, universities, and engineering firms. The funding for this research effort comes primarily from the Subscription Program, through which water utilities subscribe to the research program and make an annual payment proportionate to the volume of water they deliver and consultants and manufacturers subscribe based on their annual billings. The program offers a cost-effective and fair method for funding research in the public interest.

A broad spectrum of water supply issues is addressed by the foundation's research agenda: resources, treatment and operations, distribution and storage, water quality and analysis, toxicology, economics, and management. The ultimate purpose of the coordinated effort is to assist water suppliers to provide the highest possible quality of water economically and reliably. The true benefits are realized when the results are implemented at the utility level. The Foundation's Trustees are pleased to offer this publication as a contribution toward that end.

Edmund G. Archuleta, P.E.
Chair, Board of Trustees
Awwa Research Foundation

James E. Manwaring, P.E.
Executive Director
Awwa Research Foundation

ACKNOWLEDGMENTS

Carollo Engineers gratefully acknowledges that the Awwa Research Foundation is the joint owner of the technical information upon which this presentation is based. Carollo Engineers thanks the Foundation for their financial, technical, and administrative assistance in funding and managing the project through which this information was discovered. The comments and views detailed herein may not necessarily reflect the views of the Awwa Research Foundation, its officers, directors, affiliates, or agents.

Clearwater Research Inc. (Boise, Idaho), led by Study Director John Hetherington, Ph.D., performed all the market research work for this project. This includes collaboration on survey structure, all survey coding, execution and data analysis, and guidance in survey interpretation by Carollo Engineers personnel throughout the project.

The authors gratefully acknowledge Dr. Irwin "Mel" Suffet for his guidance in developing the average consumer sensitivity test used in this work as well as for providing review and input on interpretation of the results of all the chlorinous flavor tests performed during this work.

The authors gratefully acknowledge Dr. Joseph Coates of Coates Futurists, Inc. for development of the future forecasting work for this project.

The technical expertise of Gary Burlingame (Philadelphia Water Department) and Andrea Dietrich (Virginia Polytechnic Institute and State University, Blacksburg, Virginia) were greatly appreciated throughout the project, as well as the support of Carine Manilève (Carollo Engineers) and Marla Johnson (Philadelphia Water Department).

The authors of this report are indebted to the following water utilities and individuals for their support, cooperation and participation in this project:

- Aquarion Water Company, Monroe, Connecticut - John Herlihy
- Chicago Water Department - Eileen Flanagan, Alan Stark and Joseph Vencill
- Corpus Christi Water Utility, Corpus Christi, Texas - Eduardo Garaña
- East Bay Municipal Water Utility - Cheryl Farr and Ron Hunsinger
- Orange County Water Utility, Orlando, FL - Jacqueline Torbert and Valerie King
- Philadelphia Water Department, Philadelphia, Pennsylvania - Gary Burlingame and Marla Johnson
- South Central Connecticut Regional Water Authority, Connecticut - Cheng-nan (Mike) Weng
- United Water Idaho, Boise, Idaho - Camille Brown, Scott Rhead and William Carr
- Water District N° 1 of Johnson County, Mission, Kansas - Byron N. Johnson, Eileen Koutelas and Charles Weber

The authors also wish to acknowledge the input and guidance provided by AwwaRF Project Managers Kim Linton and Linda Reekie, and the Project Advisory Committee (PAC)—including Terri Denemy, City of Grand Rapids, Grand Rapids, Michigan, Linda Godfrey, Thames Water Utilities, Reading Berkshire, United Kingdom, Susan Hall, Kansas City Water Services Department, Kansas City, Missouri, Sarah Katz, Katz and Associates, La Jolla, California.

ACKNOWLEDGEMENTS

EXECUTIVE SUMMARY

Growing consumer dissatisfaction with water quality has contributed to the increased purchasing of point-of-use and point-of-entry (POU/POE) treatment devices and bottled water (BW); With growing privatization prospects and competition from alternative drinking water sources and treatment providers, utilities are increasingly concerned about consumer dissatisfaction and loss of confidence in the drinking water quality.

While negative media attention often creates the perception that the public is very skeptical about their drinking water safety, historical data suggests that the leading cause of consumer dissatisfaction with tap water quality involves perception of chlorinous taste and odor (Suffet et al. 1995).

While this preliminary information provides hints as to what consumer perception of chlorinous flavors are, a study was needed that incorporated all of the various aspects of public perception of tap water chlorinous flavor, both objective and subjective, to accurately quantify public perception of chlorinous flavors.

OBJECTIVES

Though average American tap water quality is very good, it is evident that an increasing fraction of the national population is turning to tap water alternatives (TWAs) as their primary drinking water source. A continuation of this trend could have serious consequences for drinking water utilities across the nation. A primary source of consumer dissatisfaction with tap water quality appears to be taste and odor, particularly chlorinous flavors. However, the true effect of chlorinous flavors on consumer purchasing choices is not well understood. Only by identifying and understanding the various driving forces behind this shift in water consumption trends can utilities begin to develop a strategy to compete effectively in the long-term drinking water market. Therefore, the objectives of this project were as follows:

- To measure consumer satisfaction with tap water quality.
- To investigate demographic trends in consumer satisfaction and consumption of TWAs.
- To identify the factors that cause consumers to purchase and use BW and POU/POE devices.
- To investigate whether these factors are real or falsely perceived.
- To develop a list of recommendations for water utilities that can improve consumer satisfaction and help bridge gaps between perception and reality.

RESEARCH APPROACH

The project approach focused on identifying public perceptions about chlorinous tastes and how these perceptions affected individuals' decisions to use tap water alternatives. The intention was to define customer attitudes and perceptions about chlorinous taste and odors in several different consumer markets in an effort to develop short- and long-term recommendations to utilities to address this issue. This was accomplished through the following tasks:

1. Summarize existing literature data on chlorinous tastes in tap water.
2. Survey chlorinous taste and odor perceptions demographically and geographically on a national and international scale. Then compare consumers' objective chlorinous taste sensitivity to their subjective opinions.
3. Analyze how drinking water sensory characteristics influence customer's perceptions of water quality and health risks.
4. Compare utility managers' perceptions of what their consumers think about tap water quality with their measured survey responses.
5. Develop guidelines and recommendations for addressing consumer satisfaction with tap water quality in general and chlorinous flavors in particular based on these results.

RESULTS

National and Market Surveys and Chlorinous Flavor Sensitivity Results

The results gathered during the national survey indicated that consumer perception of undesirable tastes and odors did not necessarily result in dissatisfaction or distrust with tap water quality. Additionally, consumer perception of chlorinous flavors was generally less unfavorable than perception of other off flavors, and consumers were generally trusting of the government and their water utility's ability to provide them with safe drinking water. However, there was still a significant portion of the population that did not trust the water industry's ability to provide safe drinking water.

In order to derive trends within the surveyed consumers, the impact of age, gender, number of children in the household, and use of tobacco products on the overall rating of satisfaction concerning taste, smell and overall water quality were assessed. Few differences and overall trends were noted. There was no difference in satisfaction or perception of safety between men and women. Older consumers tended to be more satisfied with respect to tap water aesthetics and safety than young consumers.

Differences in Satisfaction with Tap Water Quality between Free Chlorine and Combined Chlorine Markets

Average consumer satisfaction with respect to overall water quality, taste, smell, safety and healthiness is shown in Table ES.1 for the free chlorine and combined chlorine target markets, and the national market. Most consumers were satisfied with their tap water, although taste was the cause of higher consumer dissatisfaction and was often associated with a perception of tap water unhealthiness and lack of safety. There were no statistically significant differences in satisfaction between the chlorine markets as a group and the monochloramine markets as a group, though there were significant differences in satisfaction between individual markets. This was in line with survey results reported elsewhere (Mackey et al. 2003).

This survey found that the majority of consumers felt tap water was safe and was aesthetically good to drink, though there was much distrust of chlorinous tastes and odors. However the significant variation in some parameters from market to market (e.g., the fraction of consumers dissatisfied with taste ranges from 15 to 33%) highlighted the difficulty water utilities face when trying to assess their consumers' perception of water quality and what might be influencing their opinion. Local culture, water quality, water temperature, etc., all influence consumer satisfaction with tap water quality and their decisions about whether to purchase TWAs.

Table ES.1
Summary of telephone survey results for the national survey and market surveys - Satisfaction with water quality

		Free Cl_2				Monochloramine			Nat'l
		EC (%)	SE1 (%)	MW1 (%)	West (%)	SE2 (%)	WC (%)	MW2 (%)	(%)
Overall satisfaction:	Sat./Very sat.	81	66	82	74	77	79	84	72
	Dissat./Very dis.	17	33	17	24	22	20	15	26
Taste:	Sat./Very sat.	71	65	78	67	68	75	84	66
	Dissat./Very dis.	28	34	23	32	31	24	16	33
Smell:	Sat./Very sat.	85	78	89	80	84	87	89	80
	Dissat./Very dis.	14	21	9	16	15	11	9	17
Safety:	Sat./Very sat.	86	80	85	84	84	84	87	82
	Dissat./Very dis.	10	16	10	8	11	13	8	15
Health:	Sat./Very sat.	74	63	73	77	73	79	79	68
	Dissat./Very dis.	18	21	10	16	15	13	11	21

In general, the observed individual market trends reflected the national market trends - perception of no or a fresh taste resulted in a much higher perception of water safety and healthiness. Figure ES.1 illustrates the impact of perceived tap water flavors on the consumer's perception of safety for the seven drinking water markets surveyed. Chlorinous-tasting tap water was generally perceived as safer than tap water with other off flavors. However, in two markets (Midwest 2 and Southeast 1), average consumers who described their tap water as having a chlorinous taste rated the safety and healthiness of their tap water lower than average consumers who described their tap water as having a bad taste. Consumers in these two markets appeared to be particularly concerned about the health implications associated with chlorinous tastes in their drinking water.

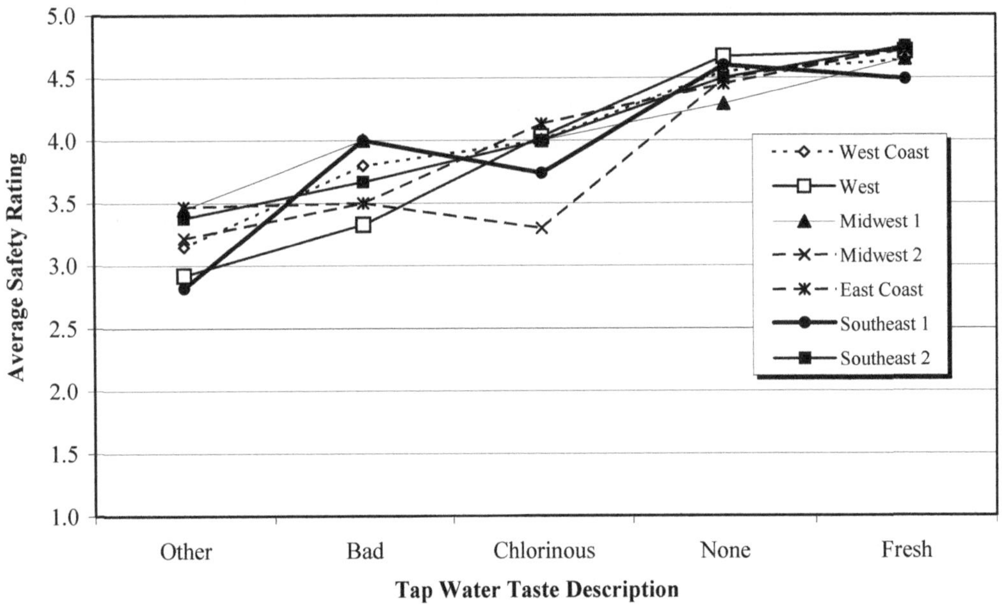

Figure ES.1 Impact of taste and odor description on tap water safety rating, observed in the national survey

While consumers may be confident overall in their tap water safety, they are not so sure about "chlorinous water." Consumers were also surveyed as to whether or not they agreed with the following statements: "In general, I believe it is safe to drink tap water;" And "I believe it is safe to drink water that has a chlorinous taste or odor" (Table ES.2). In general, about three-quarters agreed that it is safe to drink tap water, but only about one-third-to-one-half (depending upon market) thought it is safe to drink water that has a chlorinous taste and odor.

Trends describing tap water odor description vs. safety and healthiness ratings for individual markets were somewhat similar to those observed for tap water taste description vs. safety and healthiness ratings.

According to the results collected from the national and regional market surveys, poor taste was the dominant motivator that led consumers to switch from drinking tap water to drinking a tap water alternative. Since "chlorinous taste" was one of the most commonly cited taste descriptors, it was hypothesized that increased chlorinous taste would motivate consumers to switch to TWAs, however there was no obvious trend in selection of TWAs as a function of perception of chlorinous flavors in tap water.

Table ES.2
Summary of telephone survey results for the national survey and market surveys - Perception of chlorinous tap water safety

	Free Cl_2				Monochloramine			Nat'l
	EC (%)	SE1 (%)	MW1 (%)	West (%)	SE2 (%)	WC (%)	MW2 (%)	(%)
"In general, I believe it is safe to drink tap water."								
Strongly agree/Agree	72	68	74	83	73	83	73	73
Disagree/Strongly disagree	15	19	16	13	16	11	15	16
"I believe it is safe to drink water that has a chlorinous taste or odor."								
Strongly agree/Agree	37	34	36	53	30	35	37	30
Disagree/Strongly disagree	54	55	48	41	58	44	50	58

Consumers within free chlorine markets were more likely to describe their tap water taste and odor as chlorinous than consumers within combined chlorine markets, however, no statistical difference was observed between free chlorine market consumers and combined chlorine market consumers in terms of satisfaction with household tap water aesthetics or in terms of confidence that their tap water meets or exceeds all drinking water standards. Consistent with these data, it was found that 48.9 and 48.1% of combined and free chlorine market consumers, respectively, primarily drink tap water while at home. That is, there was no difference in TWA use between consumers in markets that used free chlorine and markets that used combined chlorine. Therefore, while the use of free chlorine as a distribution system disinfectant might have caused increased consumer detection of chlorinous taste and odor, it did not affect average consumer satisfaction with tap water quality nor did it account for trends in TWA usage. While some fraction of consumers will choose TWAs due to chlorinous flavor in the water, there are many factors that contribute to the growing consumption of TWAs over tap water by American consumers.

Comparison between Customer Opinions and Sensitivity

The chlorinous flavor sensitivities measured during the taste tests were associated with the consumers' perceptions (reported in the survey) and the average water quality ratings. The results obtained in the free chlorine markets can be summarized as follows:

- 46% of the tested panelists had a threshold above the residual.
- 34% of the tested panelists reported a chlorinous flavor in their water, 57% of whom had thresholds below the highest free chlorine residual in their market (at the entrance to the distribution system).
- Of those who perceive a chlorinous odor (much lower threshold), 2/3 also reported a chlorinous taste.

- There was no significant difference in tap water perception of aesthetics or safety/health between those consumers with threshold below and above the chlorine residual.
- There was no significant difference in perception of chlorinous flavors between panelists with a low threshold sensitivity and those with a high threshold sensitivity - there was no correlation between perception of chlorinous flavor and measured ability to detect it at levels commonly found in tap water.
- As with the market survey consumers, taste test panelists who perceived no/fresh/good tap water flavors were more satisfied with their water quality at all levels, regardless of their threshold sensitivity.

Regarding the monochloramine markets, a similar approach to the chlorine markets was taken, in terms of comparing consumer sensitivities with flavor perception. The results obtained are summarized below:

The great majority of the panelists (87%) were not sensitive to monochloramine residuals of 2.5 mg/L Cl_2. This assessment did not include the panelists pertaining to the MW2 market. Due to consumer recruiting difficulties, utility personnel were used for the average consumer taste tests, and therefore, were ineligible to complete the survey. However, including their results would not have made a great impact in what affects sensitivity to chloramine residual (85% instead of 87% would have a threshold above 2.5 mg/L Cl_2).

- As in the free chlorine markets, there were no differences in consumer perception of tap water aesthetics and safety/health between panelists with threshold sensitivities above and threshold sensitivities below 2.5 mg/L Cl_2.
- As with the market survey consumers, taste test panelists who did not perceive any "off" or bad flavors (including chlorinous flavors) were more satisfied regardless of their threshold sensitivity.
- There was no significant difference in perception of chlorinous flavor in tap water between the high threshold and low threshold (who, presumably, are much more likely to taste it) panelists - there was no correlation between perception of chlorinous flavor and measured ability to detect it at levels commonly found in tap water;

These results imply that consumers often do not accurately identify chlorinous tastes and/or odors they perceive in their tap water. It is speculated that consumers may misidentify some objectionable odors as "chlorinous" because that is one drinking water constituent most people know is present in the water.

No correlation was found between tap water hardness, alkalinity, or free chlorine residual concentration with threshold sensitivity to either free chlorine or monochloramine. Age and smoking habits were also examined in regards to chlorine sensitivity levels, but were found to be insignificant. Differences in satisfaction levels between the markets were insignificant. Using chlorine sensitivity to categorize consumer's satisfaction with various water quality features was examined per respondent. Initially, it had been hypothesized that consumers with lower levels of sensitivity might be more dissatisfied with various aspects of water quality, especially taste and smell, but this was not the case.

Sensitivity to Chlorinous Flavors – Comparison Between Trained and Untrained Panelists

Although some untrained panelists had sensitivities as low as the trained panelists, on average they were less sensitive. Figure ES.2 shows a comparison of the threshold sensitivity to free chlorine for untrained American panelists and trained panelists. Similar results were obtained with monochloramine. When testing chlorine dioxide, no difference was observed between untrained French panelists and trained panelists.

Regarding the relative sensitivity of the U.S. and French untrained panels to free chlorine, the difference between the two panels was mainly a shift in the flavor perception trend (Figure ES-3). The two groups had the same heterogeneity in the answers of both panels, because they were both untrained. Similar trends were found with monochloramine. The difference in sensitivity between panels is likely due to differences in chlorination residual concentration practices in the two countries (0.1 - 0.2 mg/L Cl_2 in France versus 1.0 mg/L Cl_2 in the U.S.).

Untrained consumers were able to distinguish between different flavor types (chlorine, medicinal or chemical) but were not able to recognize the descriptor. There was no impact of temperature and TDS concentration on chlorinous flavors perception of both trained and untrained panelists. The impact of earthy/musty tastes on chlorine perception was addressed by the trained panelists. The presence of a background musty taste in the chlorine dilution solution induced a shift in the perception of the chlorinous flavor - there was a significant increase of the chlorine threshold sensitivity (0.05 to 0.1 mg/L Cl_2). The increase of the musty taste background induced an antagonistic shift of the perception of chlorine flavor. A linear relationship was found between the musty flavor threshold and the chlorine taste threshold.

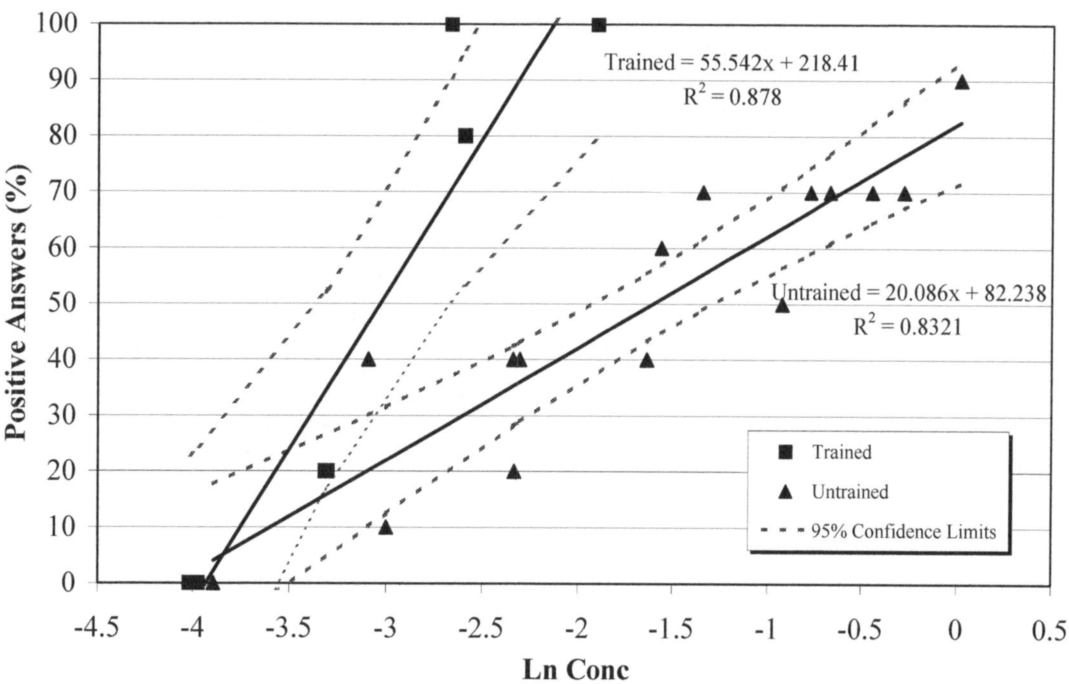

Figure ES.2 Comparison of chlorine perception for the U.S. untrained versus trained panel

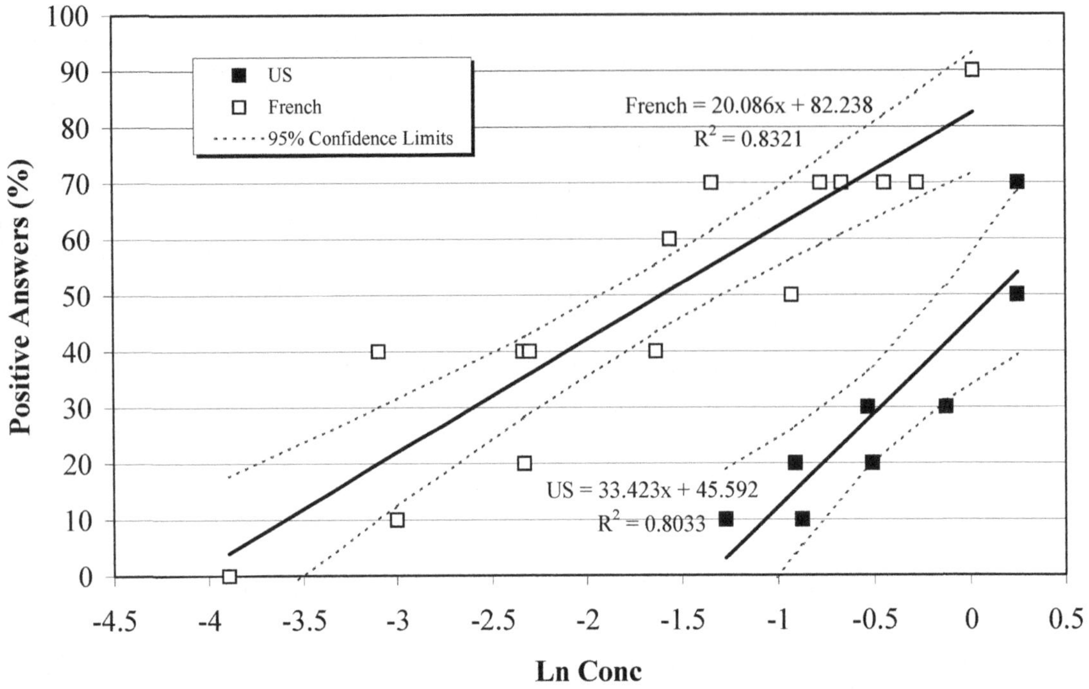

Figure ES.3 Comparison of chlorine perception between French and U.S. untrained panel

Utility Manager's Survey

The presence of chlorinous taste and odor in tap water was viewed by most managers to be an insignificant problem. Approximately 75% of all U.S. and international managers believed that consumers feel safe when they detect a chlorinous T&O in their tap water, although on average, 43% of consumers (ranging from 38 to 53%, depending on the specific market) associated chlorinous flavors with lack of safety. Most U.S. managers (53%) believed that consumers understood the benefits of chlorine, but at the same time realized its aesthetic problems need to be mitigated by taste and odor control methods. International managers, who are accustomed to either very low chlorine residuals or entirely different disinfection processes (e.g., ozone, ultraviolet radiation), did not express as much concern for chlorinous taste and odor-related issues. This might be due to their limited use of chlorine as a primary disinfectant and greater tendency to use more high tech processes to remove natural organic matter and other potentially odor-causing constituents from the water.

Recommendations

It has been documented here that:

- Consumers generally have a negative opinion of chlorinous tastes and odors in their drinking water.
- Perception of this flavor leads to lower satisfaction with tap water flavor, healthiness and safety.

- Many consumers may not be sensitive to chlorinous flavors and in many cases this flavor may be misidentified.

These facts indicate that consumer satisfaction can be increased by two methods: 1) Improving customer confidence in tap water quality and the safety of chlorinated tap water; 2) Reducing chlorinous flavors and thereby reducing their detection.

Public Education Campaigns

In many ways, it can be argued that there is little difference between tap water, bottled water and filtered water in the marketplace. They are all complementary products. Advertising and "selling" public drinking water may be a new concept for water utilities, but to remain in good standing and trust, new approaches and tactics need to be taken. Simple customer education efforts include:

- Educate the public about source water protection and drinking water treatment. Chlorination provides important public health protection from microbial pathogens. Chlorination is practiced very safely in the United States, though many of its citizens may not understand all the issues and health data associated with the presence of chlorine in their drinking water.
- Modify Consumer Confidence Reports (CCRs)/Annual Water Quality Reports
  - Make chlorine and disinfection byproducts safety and health issues a focus. Specifically address consumer concerns about the impact of chlorine on their health.
  - Use a marketing approach when designing the report, don't just provide the required tables in an unguided mass transfer of data.
  - Advertise or publish them in the local paper, on television, billboards or the radio to increase customer awareness.
  - Include a Frequently Asked Questions (FAQ) section to address common customer complaints.
- Have a public relations team and response plan in place to deal with negative "events," be it a boil water notice, a negative media report, or a chemical spill in a nearby area-combat negative press with positive press.

Minimize the Presence of Chlorinous Taste and Odor

The data developed in this report strongly suggests that average consumers are very insensitive to monochloramine. This implies that switching from free chlorine to monochloramine should solve chlorinous flavor taste and odor problems. However, there was no statistically significant difference in satisfaction with tap water between the free chlorine and monochloramine markets.

It is not expected that switching to monochloramine would solve the problem of customer perception of chlorinous tastes and odors, but it would help utilities eliminate one source of consumer taste and odor perception and would enable the utility to more accurately target the true sources of their taste and odor problem. However, it should be cautioned that switching to monochloramine is not a trivial decision and comes with its own set of operational challenges and potential problems.

FUTURE RESEARCH

Often times, research raises more questions than it answers. For this reason, an exhaustive appendix has been included with this report, which can be used beyond the analyses presented here, to further examine the area of consumer perception of chlorinous flavors and water quality in general. Specifically, an Excel™ data file has been included for those wishing to run correlations and determine relationships not presented in this report.

In addition, there are other aspects of consumer sensitivity to chlorinous flavors that warrant further research, these include the following investigations:

1. Compare individuals' ratings of the acceptability of a range of chlorinated samples with their threshold sensitivity.
2. In areas where the utility is planning to switch from a free chlorine residual to a monochloramine residual, survey consumer population in that area about perception of chlorinous flavor and overall water quality before the switch, right after the switch, and a year later.
3. Test the sensitivity levels of consumers to both free chlorine and monochloramine to see if there are differences in sensitivity between markets - i.e., are consumers in free chlorine markets more sensitive to monochloramine than consumers in monochloramine markets and visa versa?

CHAPTER 1
INTRODUCTION

BACKGROUND

Growing consumer dissatisfaction with water quality has contributed to the increased purchase of point-of-use and point-of-entry (POU/POE) treatment devices and bottled water. As of 2001, nearly 2 million POU/POE devices were being purchased annually in the U.S. and Canada (Doulton Water Filters 2001). The bottled water market is growing at a similarly high rate. U.S. sales totaled 3.4 billion gallons in 1999, and 5.7 billion in 2000, a growth rate of almost 10 % (Best Cellars 1998).

With growing privatization prospects and competition from alternative drinking water sources and treatment providers, utilities are increasingly concerned about consumer dissatisfaction and the resultant increased purchase of tap water alternative products. This growing concern is not associated with loss of revenue for the utilities, but with the loss of consumer confidence in the drinking water quality.

While negative media attention often creates the perception that the public is very skeptical about their drinking water safety, historical data suggests that the leading cause of consumer dissatisfaction with tap water quality involves perception of chlorinous taste and odor (Suffet et al. 1995).

The taste descriptors typically used by dissatisfied tap water consumers include "chlorinous," "earthy-musty," and "medicinal." There are indications that the perception threshold of a given chlorine species is influenced by the overall quality of the water. As an example, masking effects between chlorine and musty-smelling compounds tend to increase the odor thresholds for both types of descriptors when simultaneously present in a given sample. These results may imply that decreasing the natural organic matter (NOM) content of the water could likewise decrease the odor thresholds of chlorinous disinfectant.

It might seem that the easiest way to fix a chlorinous taste and odor problem is to lower the free chlorine or chloramine residual. However, this raises concerns about microbial contamination and regrowth within the distribution system. Protection of public health and delivery of aesthetically acceptable water to consumers requires that water quality be maintained to the consumer's tap. Water quality deteriorates as it travels through the distribution system. In every potable water supply, microbiological activity is invariably present in the distribution system. Aesthetically pleasing water quality can rapidly develop objectionable taste and odors from the metabolic byproducts of certain bacteria and through the interaction of chlorine with microorganisms and biofilms. Biological activity can increase corrosion rates and affect the water quality through increased color and turbidity.

For instance, ozonation converts recalcitrant NOM to biodegradable organic matter (BOM) and increases the substrate available for microbiological growth. However, coagulation, biological filtration, activated carbon adsorption, nanofiltration, and reverse osmosis reduce BOM levels. The amount of BOM present indicates the tendency of a water to promote microbial growth. Other factors impacting regrowth include detention time in the system, corrosion control, and the level of residual disinfectants (Cushing and Crozes 1999).

Further complicating this picture are the differences in consumer sensitivity from region to region. The typical potable water free chlorine residual in France is 0.3 mg/L or less. In the United States, it is on the order of 1.0 mg/L, approximately three times higher. Due to a historically low exposure to chlorine, French consumers have developed a rather high sensitivity to chlorinous taste and odors and often tend to react and complain when the free chlorine residual exceeds 0.15 mg/L. The ongoing exposure level of the consumer as well as the actual concentration in the water effects consumer sensitivity. Utilities need to understand cultural and regional perceptions of chlorinous flavors in drinking water if they are to improve customer satisfaction in this regard.

PROBLEM STATEMENT

While this preliminary information provides hints as to what consumer perception of chlorinous flavors are, a study was needed that incorporated all of the various aspects of public perception of tap water chlorinous flavor, both objective and subjective, to accurately quantify public perception of chlorinous flavors. Questions to answer included:

- What is public perception of tap water chlorinous flavors?
  - Does a significant fraction of the public perceive chlorinous flavor in their water? Does it affect their level of satisfaction with water quality?
  - What is their sensitivity to chlorinous flavor?
  - Are more sensitive individuals less satisfied?
- At what point do chlorines bother people enough for them to remedy the problem on their own?
- How does chlorine taste affect public perception of health risk associated with drinking water?
- To what degree are other aspects of the water quality impacting chlorinous perception of total dissolved solids (TDS), hardness, color, other taste and odors descriptors?
- Do Consumer Confidence Reports have a comforting or confusing result, exacerbating the consumer's skepticism?
- How well do we understand consumer perception of tap water quality? Are we on-target? Or are we way off-base?
- What are the appropriate changes for utilities to make to improve customer satisfaction? More sophisticated, centralized treatment? POU/POE devices operated by utilities? Reduce the disinfectant residual (and, thus, reduction of the disinfectant demand of the water)?

OBJECTIVES

Though average American tap water quality is very good, it is evident that an increasing fraction of the national population is turning to tap water alternatives as their primary drinking water source. A primary source of consumer dissatisfaction with tap water quality appears to be taste and odor, particularly chlorinous flavors. However, the true effect of chlorinous flavors on consumer purchasing choices is not well understood. Only by exploring the various objective and subjective components of public perception of tap water chlorinous flavor can utilities understand: 1) how chlorinous flavors influence consumer satisfaction with tap water quality; and 2) if and how utilities should address chlorinous flavors in their own finished waters.

Therefore, the objective of this project was to perform a fact-finding study aimed at understanding public perceptions about chlorinous tastes, tap water quality and how they impact customer choices with respect to tap water alternatives such as bottled water and POU/POEs. A wide range of sample populations in the U.S. and abroad were studied to incorporate the interrelated influences of region, culture, and finished water quality.

RESEARCH APPROACH

The project approach focused on identifying public perceptions about chlorinous tastes and how they affect individuals' decisions to use tap water alternatives. The intention was to define customer attitudes and perceptions about chlorinous taste and odors in several different consumer markets in an effort to develop short- and long-term recommendations to utilities to address this issue. This was accomplished through the following tasks:

1. Summarize existing literature data on chlorinous tastes in tap water. A review of previous, relevant consumer surveys and data on human perception of chlorinous flavors is presented in Chapter 2. This will provide a background understanding of existing insight on the chlorinous taste and odor issues and put the new research in context.

2. Analyze chlorinous taste and odor perceptions demographically and geographically on a national and international scale. Then compare consumers' objective chlorinous taste thresholds to their subjective opinions. A market survey was conducted to analyze customer perceptions demographically and geographically. This was done using quantitative telephone interviews for two different sampling frames in the United States. The first sampling frame included randomly-selected households from seven specific utility markets; the second sampling frame was comprised of households randomly-sampled across the country. This allowed collection of both general data and detailed data on specific regions (markets survey). It is important to understand global trends and the effects of specific water qualities and biasing events such as waterborne disease outbreaks and environmental issues. A two-pronged approach provides both a general characterization of customer attitudes/perceptions in the United States, and more detailed information on utility customers in specific markets.

   Sensitivity of individual consumers may significantly influence his/her perception of tap water quality. Threshold sensitivity taste tests (in addition to the consumer survey) were given to average consumers in each of the seven target markets to assess both the individual's taste threshold and degree of satisfaction with their tap water quality. Use of seven different markets with seven different water qualities allowed investigation of the impact of different water quality matrices on consumer sensitivity and opinion of tap water quality.

   The methodology is presented in Chapter 3 and the results of the national survey and each utility market are summarized in Chapter 4. Comparison of subjective and objective perceptions of chlorinous flavors are presented in Chapter 5.

3. Analyze how these sensory characteristics influence customer's perceptions of water quality and health risks. Trained and untrained panelists were used to evaluate human sensitivity based on the panelists' skill level. Investigations included both sensitivity to the chlorine residual present, ability to identify the flavor and the effect of background water quality. These experiments are presented in Chapter 6.

   Ultimately, the goal was to compare the threshold values presented in the literature with those of the participants in this study. Some specific objectives included: 1) identify the taste threshold of oxidants used in water disinfection (e.g., chlorine, monochloramine and chlorine dioxide); 2) compare the sensitivity of the trained versus untrained panel; 3) ensure that "chlorinous" flavors were correctly identified by untrained panelists; and 4) investigate the antagonism/synergism effects due to background water quality characteristics.

4. Identify how well utility managers understand their customers' perceptions of water quality, particularly chlorinous flavors. One-on-one surveys were conducted to assess how well utility managers knew their customers, and how this type of information could be used in the future in developing PR projects that benefit both utilities and consumers. These results are summarized in Chapter 7. Managers were surveyed on both a national (U.S.) and international (Europe and Australia) level.

5. Develop recommendations. The results from these various investigations are summarized in Chapter 8 and used to develop some guidelines and recommendations for addressing consumer satisfaction with tap water quality in general and chlorinous flavors in particular.

CHAPTER 2
IMPACT OF CHLORINOUS FLAVORS ON CONSUMER SATISFACTION

INTRODUCTION

Consumers commonly cite the presence of chlorinous flavors in their tap water as the main reason of dissatisfaction forcing them to look for alternatives (e.g., bottled and filtered water, POU/POE devices). Surveys are one of the tools that enable consumer perception to be known and associated with the parameters, such as taste and odor, or safety, enabling the identification of consumer concerns and helping to define strategies that contribute to increase their satisfaction.

Chlorinous flavors can be caused by the presence of free or combined chlorine products. However, some consumers may associate unknown flavors with the presence of chlorine and therefore report them as chlorinous. It is therefore, important to understand consumer sensitivity to chlorinous flavors, since mitigation at the plant and distribution system can only be successful if the problem is correctly identified.

OBJECTIVE

The goal of this chapter is to review previous relevant consumer surveys and to summarize existing data on human perception of chlorinous tastes.

METHODOLOGY

In order to better understand consumer attitudes with respect to perception of chlorinous flavors, such as tolerance threshold and accurate flavor perception in the tap water, a literature review was performed to gather historical data on the identification of compounds associated with chlorinous flavors, consumer sensitivity to chlorinous flavors, and on existing consumer surveys that help understand consumers' attitude towards chlorinous flavors and water quality in general.

CONSUMER SURVEYS

A better understanding of consumer attitudes towards tap water is achieved by means of surveying a population. Depending on their specific objectives, surveys can be deployed at national or local levels. They should be designed in such way that will be easy to the consumers to relate to the questions and give reliable answers, and enable the surveyor to use the information to improve solutions to the existing problems. Same surveys are described in detail below.

CIEau Survey, "The French and Water 2000"

The CIEau (Water Information Center, www.cieau.com) is a non-profit French association created in 1995 to inform the public on all aspects of domestic water using various communications tools (e.g., surveys, TV programs, colloquia, exhibits, communication campaigns, the internet). Its members are local and national water treatment plant operators. Each year, CIEau has assessed consumer satisfaction with tap water via a national survey.

In 2002, a national survey was conducted, in which 2,187 adults (> 18 years old) were interviewed in person. Overall, the results of the survey showed that three-quarters of the population was satisfied with their water quality and the majority was satisfied with their local water utility service (Figure 2.1). This survey has been conducted since 1996, and no significant shifts in customer satisfaction have been reported in the recent past. Figure 2.2 illustrates consumer attitudes towards drinking water. Although 72% claimed to be satisfied with their drinking water quality, only 52% drank it at all. Additionally, only 21% of the sampled population consumed tap water regularly.

This survey further characterized the causes of consumer dissatisfaction by grouping them into five categories: presence of chlorine, "bad" taste, pollution, hardness, and the cost (Figure 2.3). Depending on the use of the water, different problems were mentioned. Taste problems were a cause of concern for drinking water, whereas hardness was a determining factor for dissatisfaction with tap water for laundry and dishwashing. Overall, French consumers were pleased with their water quality (Figure 2.4). However, the degree of satisfaction depended on the use made of the water. In fact, drinking water quality was characterized with the lowest average satisfaction. Organoleptic criteria were commonly cited as the reason for concern and dissatisfaction.

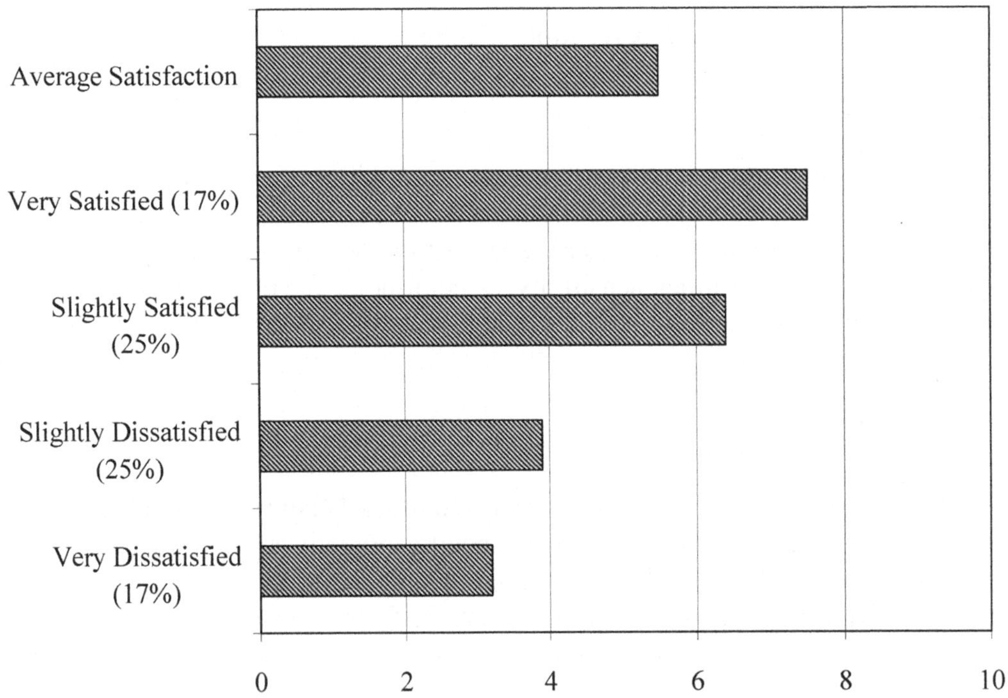

Figure 2.1 Consumer satisfaction with local water utility services (on a scale of 0 to 10)

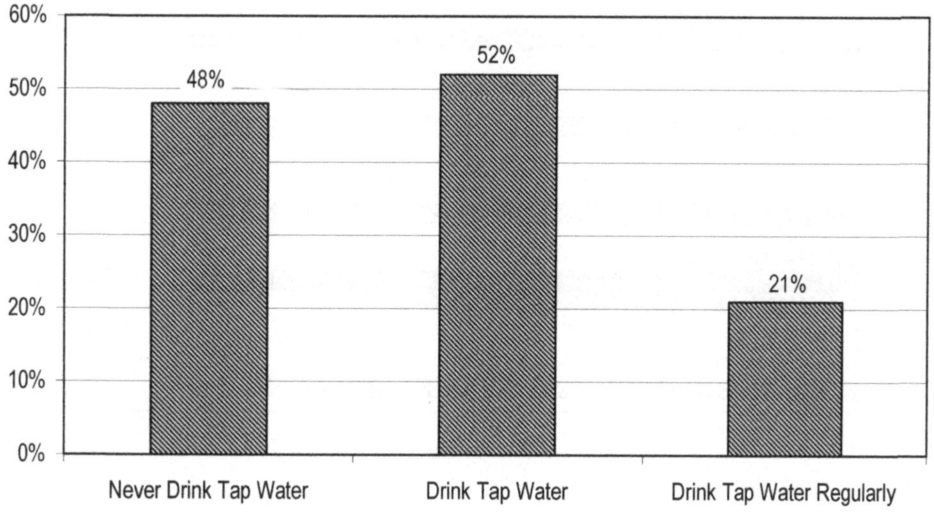

Figure 2.2 Consumer attitudes towards drinking water

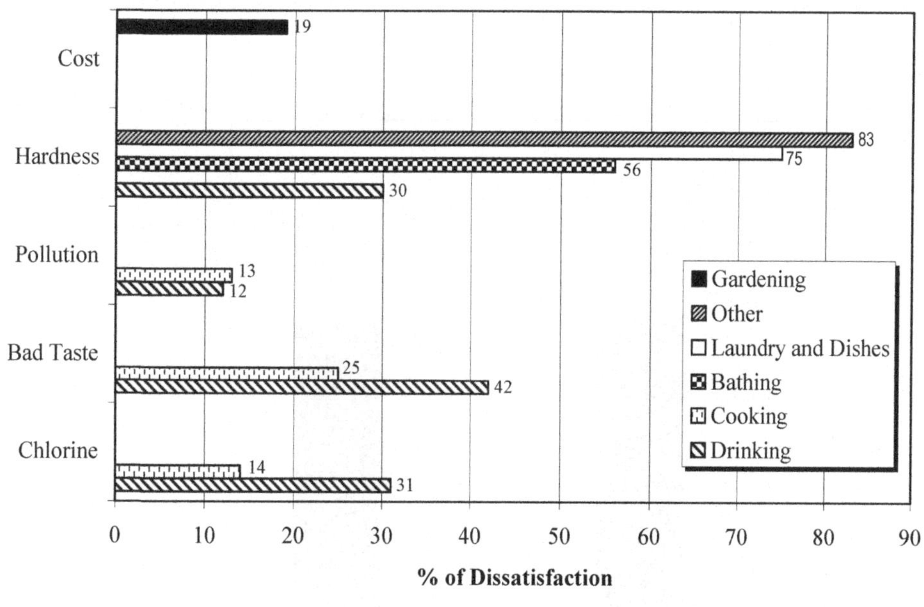

Figure 2.3 Causes of dissatisfaction related to tap water quality

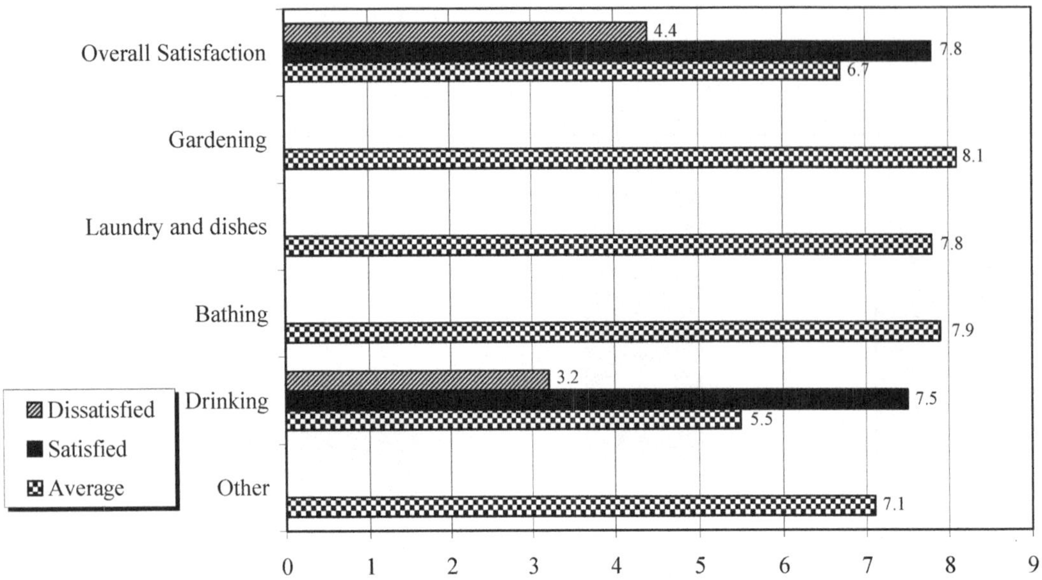

Figure 2.4 Degree of consumer satisfaction associated with various water uses

Figure 2.5 shows the relative consumption of tap and bottled water by the population. A small decrease in the number of tap water drinkers was observed with respect to the survey of the previous year.

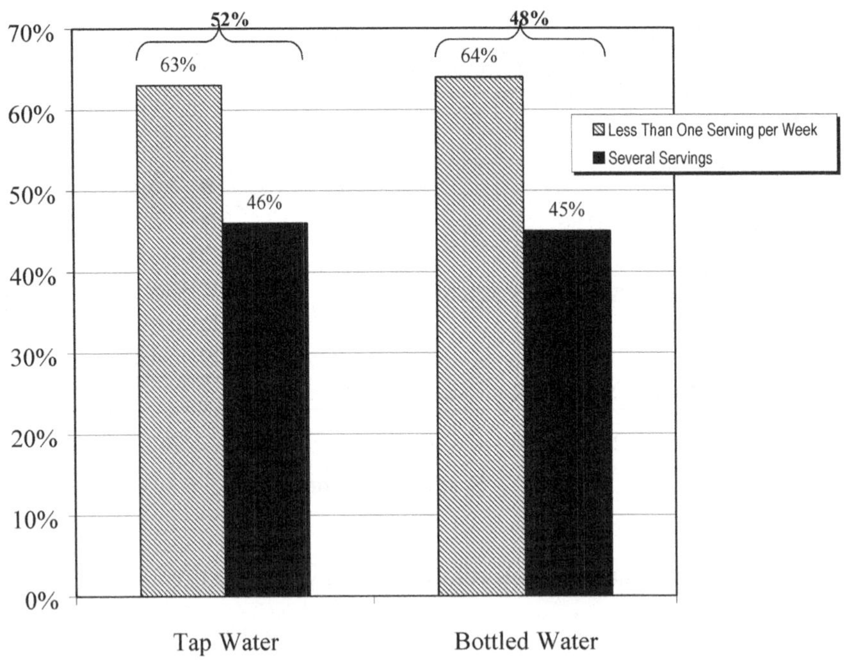

Figure 2.5 Tap water and bottled water consumption habits

United States Consumer Perception – A Comparison of Seven Surveys

Torobin, M.G. et al. (1999) summarized the results of several consumer surveys conducted in the United States in order to evaluate the importance of the aesthetic qualities of tap water on consumer perceptions with respect to its safety. Two approaches were taken. In the first approach, results of seven consumer surveys were compared to determine whether there was a systematic relationship between tap water aesthetics and perceived safety. The results demonstrated that fair/poor aesthetics were associated with unsafe water quality. Consumers reporting poor tap water aesthetics also tended to have a poor opinion of the safety of their tap water.

The second approach entailed a detailed examination of data from a telephone survey conducted with 935 consumers in Southern California. This study reported that the three most important factors for predicting consumer perception regarding tap water safety were: 1) aesthetic quality, 2) consumer confidence in the local utility, and 3) consumer confidence in utility compliance with drinking water standards. Among the aesthetic parameters, problems associated with taste were the strongest predictor of consumer dissatisfaction, leading to distrust in water quality (Figures 2.6 and 2.7). Very often, problems with taste were associated with perception of non-compliance with drinking water standards (Torobin et al. 1999). Consumers who never had taste problems with their tap water were three times more confident in their tap water than people who frequently had taste problems and were four times more inclined to believe that water quality standards are respected. Additionally, a correlation was found between aesthetics quality rating and the perceived safety rating (Figure 2.8).

Internal Ondeo Services Surveys

West Parisian Suburb - 1998

The aim of this survey was to determine the perception of tap water quality by consumers in west Parisian suburbs. Despite recurring taste problems at the outlets of certain plants, the number of consumer complaints tended to decline over time.

The survey was conducted by telephone, using a 16-minute questionnaire, on a sample of 506 people over the age of 18. Figure 2.9 summarizes the results of this survey regarding consumption habits. Taste was the descriptor most commonly cited as a reason for consumer dissatisfaction. Furthermore, one quarter of the consumers interviewed (specially those under the age of 35) reported drinking less tap water than they did in the past (Figure 2.10). This was especially true with young consumers.

Thirty-three percent of bottled water consumers reported using bottled water because they did not like the taste of their tap water. Consumer perception of chlorinous flavors and odors was relatively high (Figure 2.11), and 52% found chlorinous flavors unacceptable (Figure 2.12).

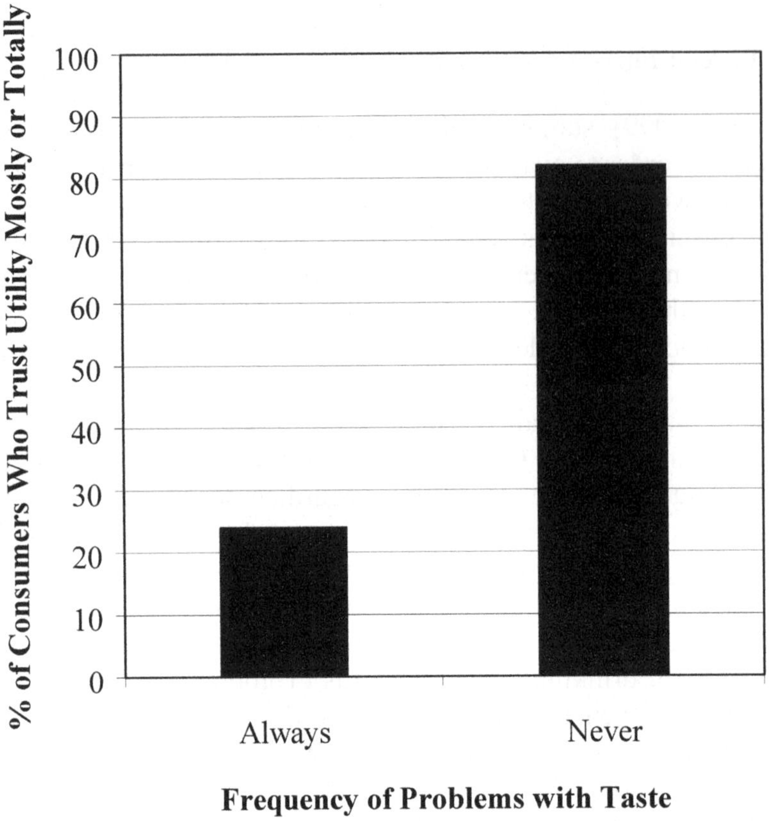

Figure 2.6 Relationship between taste episodes and trust on the utility services

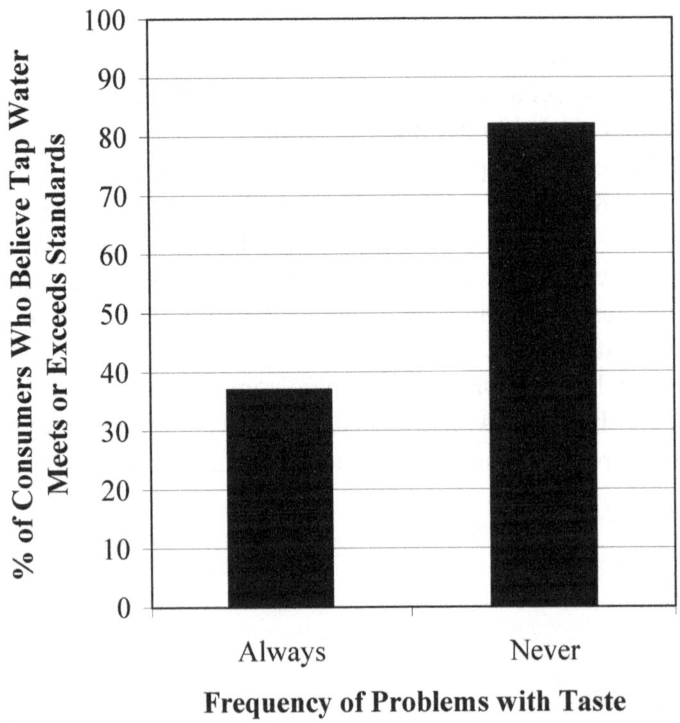

Figure 2.7 Relationship between taste episodes and water quality perception

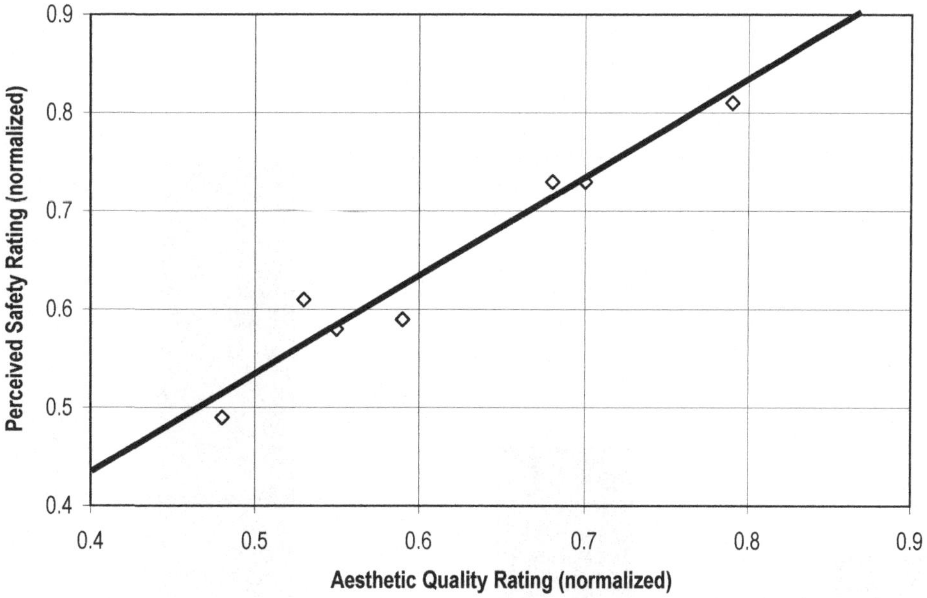

Figure 2.8 Impact of aesthetic parameters on water quality safety perception

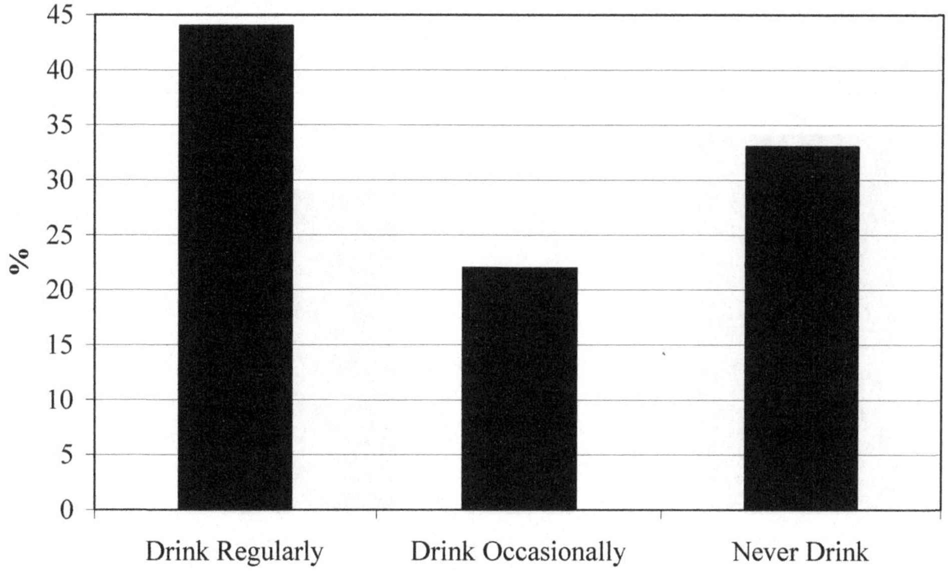

Figure 2.9 Tap water consumption habits

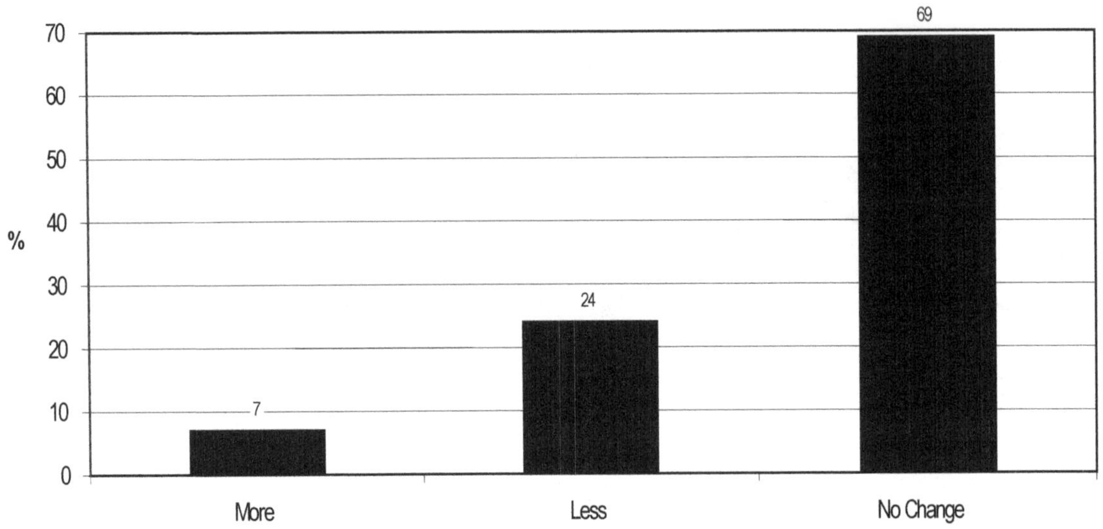

Figure 2.10 Change of tap water consumption habits within the past 2 years

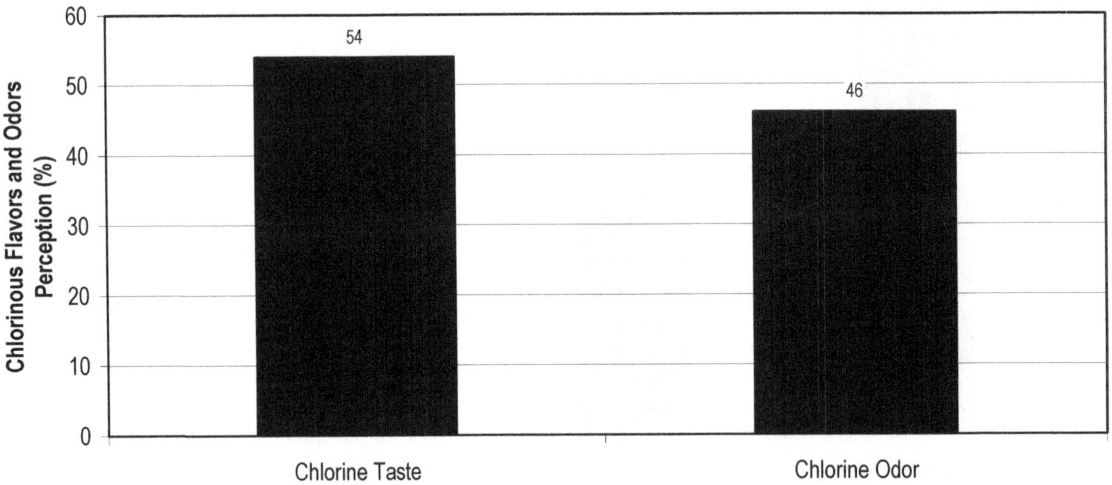

Figure 2.11 Consumer perception of chlorine flavors and odors

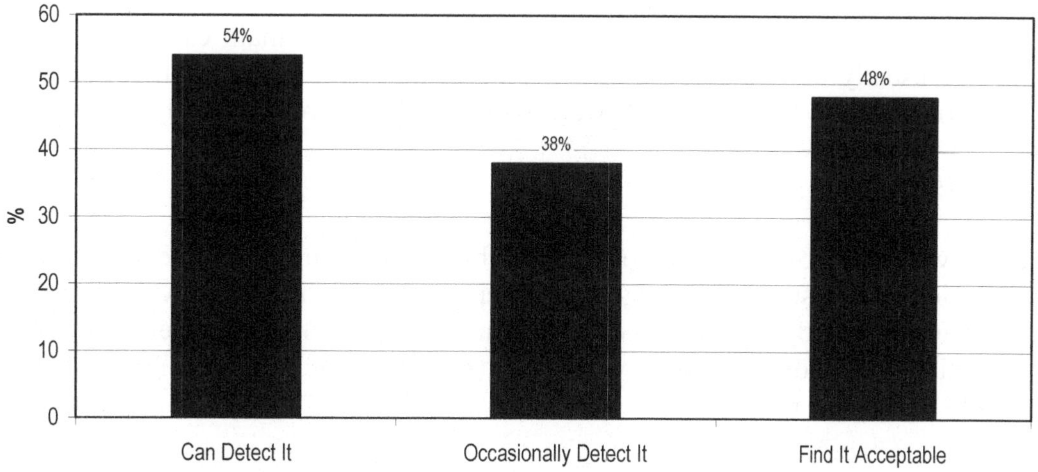

Figure 2.12 Consumer perception of chlorine taste

POU or POE devices were looked upon as a credible alternative to tap water (Figure 2.13). In general, filters were used to supply drinking water, whereas water softeners were used to protect washing machines.

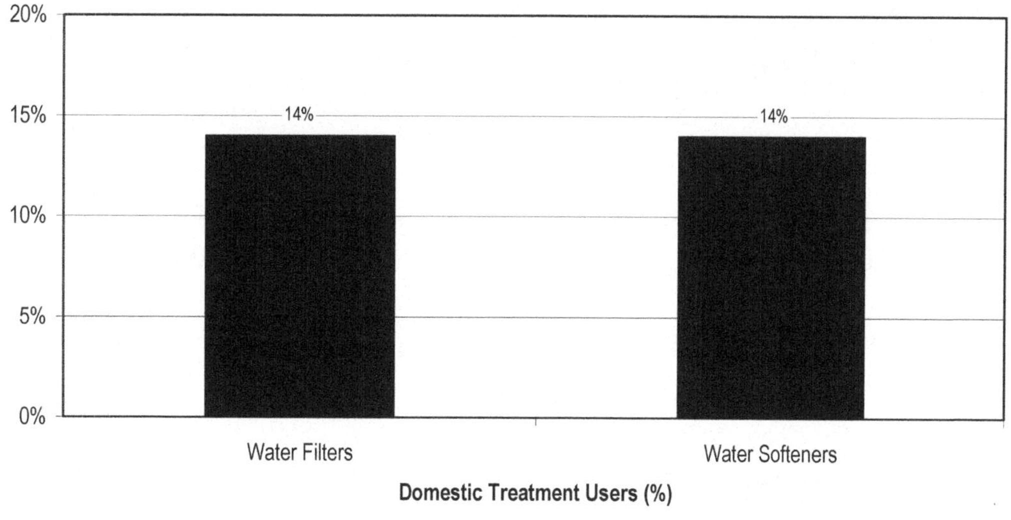

Figure 2.13 Tap water alternatives

Central France - 1999

Another telephone survey of 400 people was performed in central France. Similar trends to those previously reported were found. Chlorinous flavors were the main cause of consumer dissatisfaction, and usually detected by the entire population, regardless of their location with respect to the treatment plant. All other trends were similar to those already reported.

Eastern France - 1998

A survey was conducted in Eastern France, by a local utility, in order to evaluate consumer satisfaction with aesthetics. Survey letters were sent to 97 different towns, to a total of 15,000 clients of which 1,200 were returned (a response rate of 8%). No correlation between consumer satisfaction and chlorine residual was found (Figure 2.14). All other findings were in accordance with those already reported.

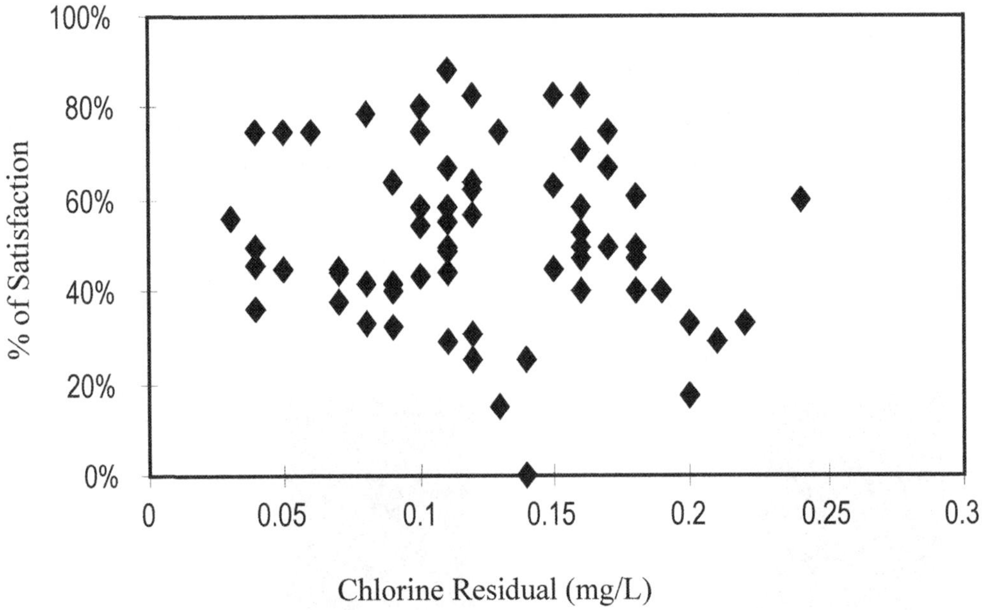

Figure 2.14 Impact of chlorine residual on consumer satisfaction

Philadelphia Water Department - 1986

This survey was conducted by Philadelphia Water Department's (PWD) Bureau of Laboratory Services and the Melicor Group. Three hundred city residents and 100 suburban residents were polled via telephone survey. Trained panelists compared bottled water, suburban water, and PWD water (with combined chlorine residuals of 0.0, 0.8, and 2.0 mg/L Cl_2 respectively).

Bottled water was preferred by 34% of the taste panelists, 36% of only PWD respondents. Suburban tap water was preferred by 28% of taste panelists, 27% of only PWD respondents. PWD tap water was preferred by 19% of the taste panelists, 18% of only PWD respondents.

Senior adults (over 60) showed no preference in taste. Young (under 40), single consumers were the most critical of tap water. Thirty-three percent of the survey respondents indicated that the presence of a chlorine residual was the primary reason they preferred tap water over bottled water.

AWWA North American Survey - 1989

This survey was conducted by the American Water Works Association (AWWA) and included 826 American and Canadian drinking water plants, of which 377 utility respondents provided useful answers about taste and odor episodes in their distribution systems (Suffet et al. 1996).

In all, 16% of operators responding considered taste and odor problems serious and 43% had at least one taste or odor episode lasting more than one week. According to these operators, the biggest taste problems were due to acidity (low pH), followed by chlorine and metallic tastes linked to the use of chloramines. Regarding the consumer, chlorinous odors were the dominant cause of complaints.

HUMAN PERCEPTION OF CHLORINOUS TASTE AND ODOR

Threshold sensitivity is the lower concentration at which a human subject can smell/taste a specific compound or class of compounds (e.g., metallic, chlorinous, musty, earthy). In this study, chlorine, chlorine dioxide, and chloramine evaluations were reviewed in depth and are summarized in Table 2.1. The wide spread in these data confirm the variability in threshold sensitivity of the subjects tested.

Table 2.1
Consumer threshold sensitivity to chlorinous flavors

Disinfectant	References	Descriptor	Conditions	Odor threshold	Taste threshold	Method
Free chlorine	Krasner & Barrett (1984)	Chlorine	25°C, pH=5	0.28 mg/L Cl_2	0.24 mg/L Cl_2	FPA*
	Krasner & Barrett (1984)	Chlorine	25°C, pH=10	0.36 mg/L Cl_2	0.30 mg/L Cl_2	FPA*
	Piriou & Perrelle (1999)	Chlorine	25°C	0.05 mg/L Cl_2	0.025 mg/L Cl_2	Threshold method
	Philadelphia Water Department (1994)	Bleach, chlorinous, sharp, burning	45°C	0.07 - 0.13 mg/L Cl_2		FPA
	Bryan et al. (1973)		40°C pH=5 pH=7 pH=9 pH not adjusted		0.075 mg/L 0.156 mg/L 0.45 mg/L 0.05 mg/L	Triangle test with statistical analysis

(continued)

Table 2.1 (continued)

Disinfectant	References	Descriptor	Conditions	Odor threshold	Taste threshold	Method
Free chlorine in DS†	Piriou & Perrelle (1999)	Chlorine	25°C		0.05-0.1 mg/L Cl_2	Threshold method
ClO_2	Hoehn et al. (1990)	Strong chlorine		0.2 mg/L	0.2-0.25 mg/L	FPA*
NH_2Cl	Krasner & Barrett (1984)	Chlorine	25°C, pH=8-9	0.65 mg/L Cl_2	0.48 mg/L Cl_2	FPA*
	Piriou & Perrelle (1999)	Chlorine			0.28 mg/L Cl_2	
	Philadelphia Water Department (1986)	Chlorinous, swimming pool, sharp	45°C, pH=10	0.13 - 0.32 mg/L		FPA
$NHCl_2$	Krasner & Barrett (1984)	Swimming pool	25°C, pH=4	0.15 mg/L Cl_2	0.13 mg/L Cl_2	FPA*
NCl_3	Suffet et al. (1995)	Geranium		0.02 mg/L		FPA*
	Philadelphia Water Department (1986)	Chlorinous, swimming pool, sharp, bleach, burning	45°C, pH=5	0.11 - 0.19		FPA
	Baylis (1935)	Sharp chlorine				

* FPA flavor profile analysis

† DS distribution system

In addition to chlorinous chemicals, disinfection by-products often produce taste and odors. For example, research indicates that some drinking water taste and odor episodes during cold periods may be due to the oxidation of urea (used as a de-icing agent) by chlorinated disinfectants (Welte and Montiel 1999, Gittelman and Luitweiler 1996), or to urea contamination of the water source that can produce trichloramine through its reaction with chlorine (Samples 1959). Chlorination of certain nitrogen compounds (e.g., amino acids, ammonium) leads to the formation of strong-smelling compounds such as aldehydes (Hrudey et al. 1989), nitriles, and organic chloramines, which can cause pronounced chlorinous tastes and odors (Kajino et al. 1999). Water treated with chlorine dioxide can react with compounds from paint to new carpeting to create "cat urine" or kerosene odors (Hoehn et al. 1990, Lykins et al. 1990, Dietrich et al. 1992). Chlorophenols, resulting from the reaction of chlorine with phenolic compounds, more prevalent in industry areas (Lin 1976, Suffet et al. 1993), can be formed in the plant and/or distribution system and impart taste and odor to the water (Bruchet 1999). Table 2.2 shows a list of chlorophenols and associated threshold sensitivity limits.

Table 2.2
Perception thresholds of halophenols

Molecule	References	Descriptor	Test conditions	Odor threshold	Taste threshold
φCl(2)	Burttschell et al. (1959)		Room temperature	0.002 mg/L	0.004 mg/L
	Schlett (1992)		Room temperature	0.00025 mg/L	
	Young et al. (1996)	Chemical	Odor at 40°C taste at 25°C	0.000088 mg/L	0.00014 mg/L
φCl(4)	Baker (1963)*			1.24 mg/L	
	Faust & Aly (1983)	Medicinal		1.24 mg/L	
	Burttschell et al. (1959)		Room temperature	0.25 mg/L	>1 mg/L
	Schlett (1992)		Room temperature	0.008 mg/L	
	White, *Handbook of Chlorination* (1972)	Medicinal		0.0005-1.2 mg/L	
	Mallevialle and Suffet (1987)			0.0005-1.2 mg/L	
	Young et al. (1996)	Chemical	Odor at 40°C taste at 25°C	0.010 mg/L	0.039 mg/L
$\varphi Cl_2(2,4)$	Burttschell et al. (1959)		Room temperature	0.002 mg/L	0.008 mg/L
	Schlett (1992)		Room temperature	0.002 mg/L	
	Baker (1963)*			0.21 mg/L	
	Faust & Aly (1983)	Medicinal		0.21 mg/L	
	White (1972)	Medicinal		0.02-0.21 mg/L	
	Young et al. (1996)				
	Young et al. (1996)	Medicinal antiseptic	Odor at 40°C Taste at 25°C	0.0054 mg/L	0.00098 mg/L
$\varphi Cl_2(2,6)$	Burttschell et al. (1959)		Room temperature	0.003 mg/L	0.002 mg/L
	Schlett (1992)		Room temperature	0.0001 mg/L	

(continued)

Table 2.2 (continued)

Molecule	References	Descriptor	Test conditions	Odor threshold	Taste threshold
	Young et al. (1996)	Medicinal antiseptic	Odor at 40°C taste at 25°C	0.0059 mg/L	0.0000062 mg/L
$\varphi Cl_3(2,4,5)$	Young et al. (1996)	Antiseptic	Odor at 40°C taste at 25°C	0.063 mg/L	0.1 mg/L
$\varphi Cl_3(2,4,6)$	CIRSEE	Medicinal	45°C	0.1 mg/L	
	Schlett (1992)		Room temperature	0.025 mg/L	
	Young et al. (1996)		Odor at 40°C taste at 25°C	0.38 mg/L minimum detected	>0.012 mg/L
φCl_5	Young et al. (1996)		Odor at 40°C taste at 25°C	0.0093 mg/L	0.008 mg/L
$\varphi Br(2)$	Whitfield (1988)	Phenol/iodine		0.00003 mg/L	
$\varphi Br(4)$	Whitfield (1988)	Phenol		0.023 mg/L	
$\varphi Br_2(2,4)$	Whitfield (1988)	Phenol		0.004 mg/L	
$\varphi Br_2(2,6)$	Whitfield (1988)	Medicinal iodoform		0.0000005 mg/L	
$\varphi Br_3(2,4,6)$	Whitfield (1988)	Iodoform		0.0006 mg/L	

* Triangle test

Thrihalomethanes (THM) is another class of compounds often associated with chlorinous flavors (i.e., medicinal, chemical). These compounds result from the reaction of chlorine with organic matter and are often present in the plant and/or distribution system. Table 2.3 shows a list of THMs and their associated threshold sensitivity limits.

Table 2.3
Perception thresholds of halomethanes

Molecules	References	Descriptor	Conditions	Odor threshold	Taste threshold
$CHCl_3$	Young et al. (1996)	Chemical antiseptic	Odor at 40°C taste at 25°C	7.5 mg/L	1.2 mg/L
	Young et al. (1996)	Medicinal		0.1 mg/L	
$CHCl_2Br$	CIRSEE	Medicinal	45°C	0.002 mg/L	

(continued)

Table 2.3 (continued)

Molecules	References	Descriptor	Conditions	Odor threshold	Taste threshold
$CHClBr_2$	Faust & Aly (1983)	Medicinal		0.3 mg/L	
$CHCl_2Br$					
$CHBr_3$	CIRSEE	Medicinal	45°C	0.005 mg/L	
CH_2ClI	CIRSEE	Medicinal	45°C	0.002 mg/L	0.005 mg/L
CH_2I_2	Suffet et al. (1995)	Medicinal		0.00002 mg/L	
CHI_3	CIRSEE	Medicinal	45°C	0.0005 mg/L	
	Moncrieff (1951)			5 ppb in the air	

Table 2.4 lists compounds often present in drinking water (either naturally or as a disinfection by-product) that are associated with chlorinous flavors.

Table 2.4
Perception thresholds of miscellaneous products

Molecule	References	Descriptor	Test conditions	Odor threshold	Taste threshold	Method
Naphth-1-ol	Baker (1963)*	Medicinal		1.29 mg/L		Triangle test
4-chloro-2-methylphenol	Young et al. (1996)	Chemical medicinal	Odor at 40°C Taste at 25°C	0.062 mg/L	0.0025 mg/L	
Benzene	Young et al. (1996)	Phenol	Odor at 40°C Taste at 25°C	0.19 mg/L		
Chloro-benzene	Young et al. (1996)	Medicinal chemical	Odor at 40°C Taste at 25°C	0.19 mg/L	0.19 mg/L	
1,3-dichloro-benzene	Young et al. (1996)	Medicinal disinfectant	Odor at 40°C Taste at 25°C	0.077 mg/L	0.19 mg/L	
Cl_3C-COOH	Ondeo Research and Technology	Perfume, chlorine	45°C	0.05 mg/L		
o-cresol	Baker (1963)*	Medicinal		0.65 mg/L		Triangle test
m-cresol	Baker (1963)*			0.68 mg/L		Triangle test
Iodocresol	CIRSEE	Medicinal, burnt	45°C	0.01 mg/L		

* Triangle test

SUMMARY

The consumer surveys showed that consumers were very sensitive to the aesthetic qualities of tap water, and to tastes and odors in particular. Of these, the problems due to chlorine or chlorinated disinfectants were the most commonly reported as the reason for consumer dissatisfaction with tap water, causing them to suspect its quality and safety, and in many cases, switch to tap water alternatives.

Disinfectants and disinfectant by-products can impart taste and odors to water. The extent to which these taste and odors can be perceived by the consumers depend on cultural habits and how the flavor tests are conducted. Although untrained consumers may confuse taste and odors, erroneously reporting a taste and/or odor problem as "chlorinous," chlorinous DBP and chlorinous taste and odors are different.

CHAPTER 3
NATIONAL AND MARKET OPINION SURVEYS AND TASTE TESTS - DEVELOPMENT AND IMPLEMENTATION

TELEPHONE SURVEY METHOD

From September 2000 to April 2002, a telephone survey of household residents over 18 years of age was conducted in seven water utility markets throughout the United States (Target Markets), as well as in a nationwide data collection effort (National Market). The survey was developed to collect general information about public perception of water quality, particularly chlorinous tastes and odors in drinking water. This section describes the methods used to develop the survey instrument, construct the sample design, conduct the telephone survey, and prepare the data for analysis.

Survey Instrument

The questionnaire explored consumer attitudes and use patterns by eliciting respondent beliefs, opinions, and behaviors related to their daily consumption of tap water (TW), bottled water (BW), and POU/POE-treated (filtered) water (FW). The questionnaire instrument was developed for the non-institutionalized general adult population of the United States with design elements specifically incorporated for administration using computer-assisted telephone interviewing (CATI). Before fielding, several survey questions were evaluated by internal and external reviewers to determine whether the wording and response categories were appropriate for the population of interest. In addition, pilot tests were conducted to ensure programming accuracy and effectiveness of interviewer training. Additional information regarding these procedures is provided below. The average length of the interview was 14 minutes. The survey was offered in English only. The content of the questionnaire is detailed in Appendix A.

Scaling Choices

Response options to all attitudinal questions were offered as a fully anchored 5-point Likert scale (i.e., "Very Satisfied," "Somewhat Satisfied," "Neither Satisfied nor Dissatisfied," "Somewhat Dissatisfied," and "Very Dissatisfied"). As many of the questions used different response categories, interviewers read all of the response options to respondents for each question. Likert-response questions were always grouped together in each section of the instrument to improve respondent consistency and reduce the reading load on the interviewer.

Questionnaire items that inquired about the frequency of actions, events, or behaviors (e.g., "Do you drink the tap water you receive in your household?") were initially asked as a simple dichotomous question (i.e., "Yes" or "No") to ensure relevancy of the question to the respondent. If the response was affirmative, respondents were then asked to report the frequency of occurrence using either specific quantities (e.g., "Once," "Two to three times," or "More than three times") or general categories (e.g., "Hardly ever," "Sometimes," or "More often"). Using this two-step approach, questionnaire skip patterns were used to channel respondents to appropriate questions, thereby tailoring the questionnaire to each respondent (e.g., chlorinous, musty/earthy, metallic).

Response items for categorical questions were developed individually for each question (e.g., "How do you describe the taste of the tap water you receive in your household?"). Items that required categorical response options fell into two general categories based on whether the response categories were read or not read to the respondent. For those questions where the response categories were read, the respondent was asked to select which category (or categories) best described his or her answer. If the order of presentation for those response lists that were read could influence a respondent's answer, then those response categories were randomly presented to the respondent. For those questions where the response categories were not read, interviewers used the pre-existing categories to code the respondent's unprompted answer to the question at hand.

Interviewers were trained to use the "Other" response category to capture responses that were either unanticipated or novel. These responses were collected verbatim as open-ended responses, and they were subsequently coded with the other response categories during preparation for data analysis. Interviewers were also trained to code non-responsive answers as either "I don't know/I don't understand" or "Refused" where appropriate.

Sample Design

The aim of sampling is to equate unknown characteristics that may influence variation in the data and to preserve the representativeness of the sample to the target population (Henry 1990). The preferred method of ensuring the credibility and applicability of research results is to employ random sampling, where every member of a target population has a known, non-zero probability of being included in the sample. For telephone-based survey research, the typical sampling plan involves random digit dialing (RDD) of household phone numbers, where information about the distribution of residential telephone numbers is used to define the sampling frame (Cochran 1977).

Description

The sample design was comprised of both nationwide and market level sample frames. All sample records were generated separately for each sample frame.

The national sample frame was specified by creating four regional areas with approximately the same population size according to the 2000 U.S. Census. Divisions between these regional areas were demarcated by the state boundaries that were included within each regional area (Figure 3.1). A total of 100 people were sampled in each of these four regions.

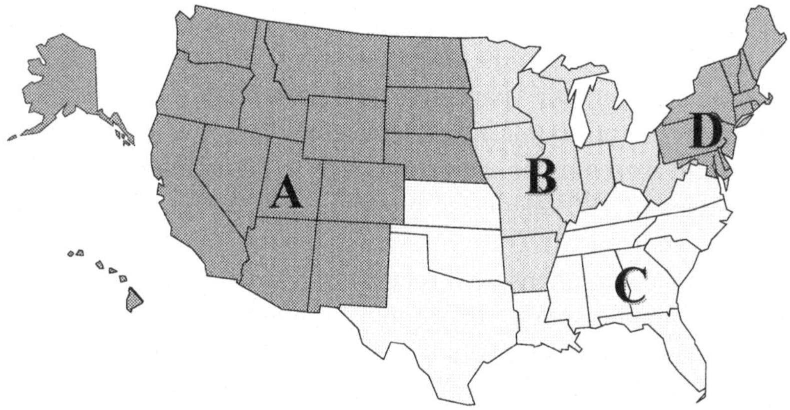

Figure 3.1 Division of states into four regions for national sample frame

The market-specific sample frame included seven strategically selected areas that were served by a single-water utility provider (Figure 3.2). These markets were geographically identified to define the sample frame for each market. A total of 200 interviews were conducted in each market.

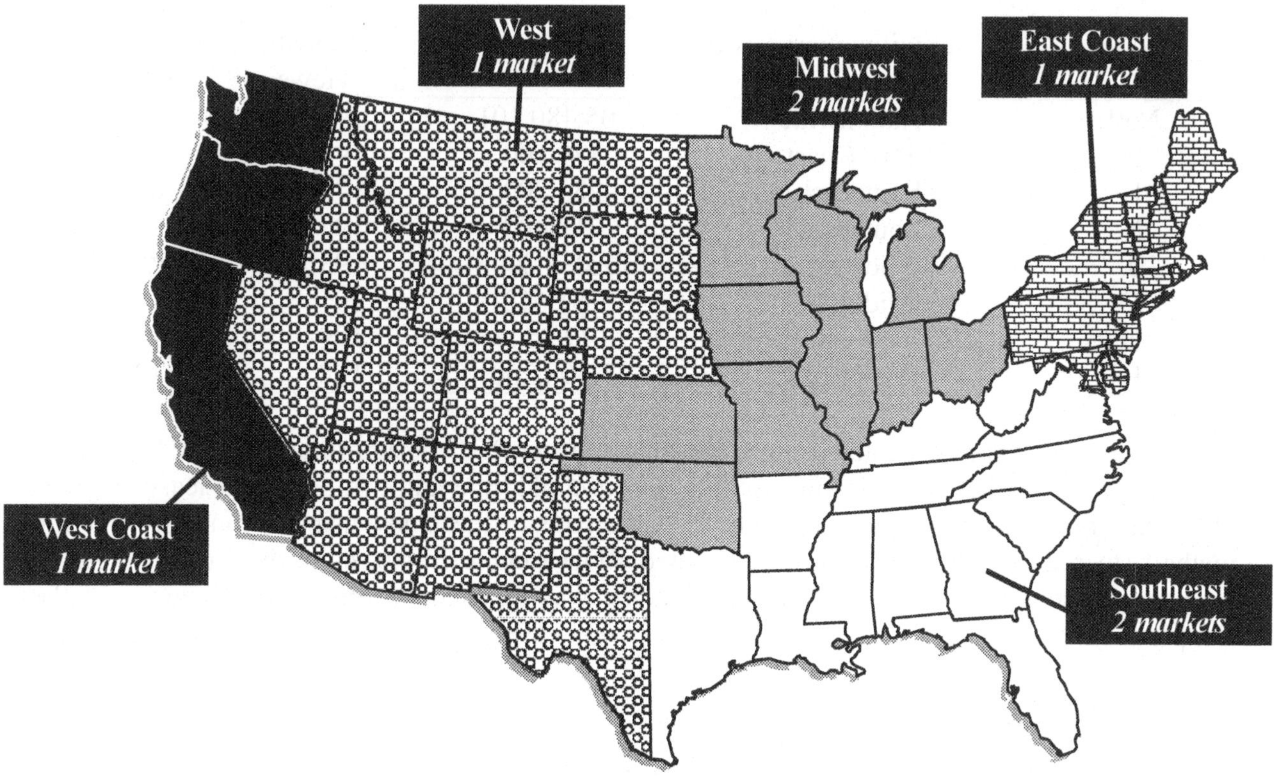

Figure 3.2 Locations of the seven target water utilities

With the exception of the West market area, a random sample of telephone numbers was generated for each water utility market area and regional area for the nationwide sample. The West market area could not be adequately specified geographically, so a complete list of residential customers—and corresponding contact information—was provided by the water utility. A random sample of households was drawn from this customer list to minimize any systematic biases introduced using this particular sampling method.

For the other six water utility market areas and the national sample, the sample frame was defined by the set of valid working banks present in each geographically defined market area. A "working bank" is a series of 100 telephone numbers defined by a particular combination of a valid 3-digit area code and a valid 3-digit exchange—the first 6 digits of a 10-digit phone number—that is found in a comprehensive directory of household telephone numbers.

Before the phone numbers were distributed to interviewers, the sample record files were purged of nonresidential telephone numbers using the IDPlus procedure provided by the sample vendor (Marketing Systems Group/GENESYS). This procedure used a combination of directory searches and line signal checks to remove non-residential telephone numbers (e.g., business or fax numbers).

A disproportionate sampling strategy was adopted to ensure comparable numbers of completed interviews were obtained for each water utility market area regardless of the population size located within each area (Table 3.1).

Table 3.1
Water utility market area population information

Code	Water utility market area	Total households	Total completed interviews	Population sampled (per million)
Natl	Nationwide	105,480,101	399	3.8
D	East Coast	379,100	192	506
L	Southeast 1	285,200	176	617
N	Southeast 2	212,400	195	918
J	Midwest 1	2,037,800	193	95
F	Midwest 2	181,612	201	1,107
B	West	77,850	199	2,556
H	West Coast	1,598,200	198	123

The survey was executed using a two-stage cluster RDD design (including the list sample from the West market area). Upon reaching a household, one adult (a person at least 18 years old) was chosen from among all adults to participate in the study using a random selection method (to reduce selection bias). A truncated RDD sample frame, comprised of all possible telephone numbers in "one-plus" working banks, was adopted to boost the efficiency of the sample.

Quota Cells

To ensure that each sample (TW and TWA drinkers) contained enough respondents, quota cells were established for each group. The sample was managed such that each quota cell contained no more than 100 of the 200 completed interviews from each market area (or 50 of the 100 for each region of the nationwide sample). A screening question (QuotaQ) was used to identify respondent's eligibility based on drinking water preference.

Although interviews were not completed with respondents if a quota cell was full, the screening information was still collected from every respondent. This information was used to project the overall incidence of TW and TWA drinkers in the population (market area). Table 3.2 presents the projected incidence of these groups in each market surveyed.

Table 3.2
TW and TWA preferences in the surveyed areas market

Type of residual	Code	Water utility market area	Total TW drinkers	Total TWA drinkers	Projected incidence of TW drinkers	Projected incidence of TWA drinkers
FC*	D	East Coast	92	119	44%	56%
FC	L	Southeast 1	73	203	26%	74%
FC	J	Midwest 1	106	87	55%	45%
FC	B	West	134	100	57%	43%
CC*	N	Southeast 2	93	118	44%	56%
CC	F	Midwest 2	171	101	63%	37%
CC	H	West Coast	98	117	46%	54%

FC Free chlorine
CC Combined chlorine

Generalizability Estimates (Projectability)

A sample size of 200 for each water utility market area and 400 for the nationwide data collection effort was determined sufficient for the accuracy of the survey results. For the nationwide survey, the actual total achieved sample size was 399. In both cases, the obtained sample size was lower than originally intended given the constraints of the fielding period and the incidence rate of tap, bottled, and filtered water drinkers. Table 3.3 summarizes the projectability estimates for this survey by group.

Table 3.3
Projectability estimates by water utility market area

Code	Water utility market area	Completed interviews	TW drinkers	TWA drinkers
D	East Coast	192 (±7%)	92	100
L	Southeast 1	176 (±8%)	76	100
N	Southeast 2	195 (±8%)	93	102
J	Midwest 1	193 (±7%)	100	93
F	Midwest 2	201 (±7%)	99	102

(continued)

Table 3.3 (continued)

Code	Water utility market area	Completed interviews	TW drinkers	TWA drinkers
B	West	199 (±7%)	100	99
H	West Coast	198 (±7%)	98	100
Natl	National Sample	399 (±5%)	206	193
Total		1753 (±2.4%)	864 (±3.3%)	889 (±3.3%)

Data Collection

This section describes the testing of the questionnaire and presents the fielding procedures adopted throughout the data collection stage of the research project.

Fielding Dates

Data was collected from September 8, 2000 through April 1, 2002. The specific fielding periods for each water utility market area are listed in Table 3.4 below.

Response Rates

The CATI system managed and monitored calling, controlled distribution of telephone numbers to interviewers, and consolidated the collected survey data. A complete listing of the algorithms used to assign the final disposition codes and the market final disposition code values are provided in Appendix B.

The response rate measures the relative success with which households sampled for the survey actually participated. For RDD samples, this is typically calculated as the percentage of households assumed to be reachable via the sampled telephone numbers that completed interviews during the field period. The higher the response rate, the lower the potential will be for nonresponse bias in the data. Since the samples were chosen randomly, a higher response rate measures that the population is well represented by the sample.

Table 3.4
Fielding dates by water utility market area

Code	Water utility market area	Start of field period	End of field period	Total N° of days in the field
D	East Coast	December 19, 2001	February 15, 2002	58
L	Southeast 1	December 19, 2001	April 1, 2002	103

(continued)

Table 3.4 (continued)

Code	Water utility market area	Start of field period	End of field period	Total N° of days in the field
N	Southeast 2	January 18, 2002	March 13, 2002	54
F	Midwest 2	March 9, 2001	June 26, 2001	109
J	Midwest 1	January 10, 2002	March 30, 2002	79
B	West	March 1, 2001	March 11, 2001	10
H	West Coast	October 25, 2001	February 11, 2002	109
Natl	National Sample	September 8, 2000	October 29, 2000	51

Two response rates developed and popularized by the Council of American Survey Research Organizations (CASRO) were used (CASRO response rate and the CASRO *Upper Bound* response rate). The CASRO rate is a more conservative estimate of the response rate, while the Upper Bound rate is a commonly used "best case" rate. Both of these response rates were calculated independently for each utility market area, and the results of this analysis are reported in Table 3.5. In addition, the refusal rate was also calculated for each utility market area.

Table 3.6 presents industry-wide survey performance statistics from the Marketing Research Association (MRA) database for all other RDD phone surveys and for all seven utility market areas and the nationwide sample. The performance rates for this study surpassed the performance of most other RDD studies in the MRA database, halving the overall refusal rate.

Table 3.5
Response, cooperation, and refusal rates by utility market area

Code	Water utility market area	CASRO response rate	Upper bound	Refusal rate
D	East Coast	10.3%	16.6%	31.0%
L	Southeast 1	6.1%	10.5%	28.9%
N	Southeast 2	13.6%	19.9%	23.5%
J	Midwest 1	9.2%	17.1%	20.9%
F	Midwest 2	7.0%	14.7%	28.7%
B	West	10.4%	32.2%	18.7%
H	West coast	11.4%	19.6%	25.9%
Natl	National sample	11.7%	18.7%	34.0%

Table 3.6
Comparison of survey performance rates

	MRA overall RDD	National sample	Target markets
Response rate	11.7%	11.7%	9.7%
Refusal rate	39.6%	34.0%	25.4%

Data Preparation

Data were converted from the CATI database and formatted for review and analysis in SPSS, a statistical analysis software package. Prior to analysis, the original survey variables and response categories were labeled, and additional variables were created for the analysis as needed. Open-ended responses were examined and cleaned for overall comprehension.

The following sections provide additional detail regarding the specific procedures employed to translate the information collected from survey respondents via the CATI system into an electronic database that could be subsequently analyzed. The protocol used to code and label the raw data will be presented first. Finally, an in-depth discussion of the weighting procedure will be elucidated.

Coding

After data collection was completed, any interviewer errors documented by data collection staff were corrected directly in the CATI database. Survey data were then exported from the CATI system into a Statistical Package for the Social Sciences (SPSS software package) format for cleaning, editing, and labeling.

A standard set of procedures was used to prepare data for review and analysis. First, each variable was provided a unique label matching the CATI question number from the survey instrument. Multiple response questions received additional subscripting to ensure that each variable label was unique. Next, each raw, labeled variable was recoded into a new variable to remove non-responsive answers (e.g., Don't Know, Refused). These recoded variables were designated using an alphabetical subscript that identifies the resultant measurement scale. A complete listing of these recoded subscripts is provided in Table 3.7.

Table 3.7
Examples of subscript labels used in coding

Label	Meaning
M	A variable which has been coded for interval level analysis. Answers such as "Don't know," "Refused," and "No answer" recoded as system missing (e.g., Q005M).
D	A variable which has been coded for use as a dichotomous (two-category) variable. Answers such as "Don't know," "Refused," and "No answer" recoded as system missing (e.g., Q005D).
T	A variable which has been coded for use as a trichotomous (three-category) variable. Answers such as "Don't know," "Refused," and "No answer" recoded as system missing (e.g., Q005T).
C	A scaled or categorical variable which has been recoded into more than three categories. Answers such as "Don't know," "Refused," and "No answer" recoded as system missing (e.g., Q005C).
A	A code for an open-ended response that has been coded into multiple categories. Answers such as "Don't know," "Refused," and "No answer" recoded as system missing.

In a separate process, open-ended questions were exported from the CATI system, and edited, cleaned, and coded in a separate electronic database before being merged with the complete SPSS database. This procedure allowed researchers the opportunity to separately examine a variety of different coding strategies before open-ended responses were aggregated with the remaining information.

Weighting

Weighting is a simple statistical adjustment (a multiplier) to each survey respondent in the data set. The purpose for weighting is usually to: 1) adjust samples to match known population parameters; or, 2) correct for disproportionate results (i.e., over- or under-representation) that occurred as a result of sample design or variations in patterns of response. Because the areas identified by the water utilities did not correspond to U.S. Census information and incidence rates of tap, bottled, and filtered water were not known beforehand, the weighting factor could only account for varying probabilities of selection. However, post-stratification weights were created for the nationwide sample using age and gender estimates from the 2000 Census.

The respondent interviews were organized for analysis in a database where each case represented data from a single interview. The cases were then weighted to reflect the sample design, to account for probabilities of selection that varied from household to household, and to minimize non-sampling error. First, a probability-of-selection weighting factor was calculated that included the different sampling fractions among the water utility market areas and the number of adults living in the household. This probability-of-selection factor was used to produce a relative weight, which could be used to calculate point estimates of population characteristics and variance estimates for statistical tests.

Data Analysis

The initial analyses involved frequency tables and descriptive statistics (e.g., mean, standard deviation) to examine and characterize the distribution of responses for each variable. These descriptive statistics also guided the subsequent analyses.

The next step in the analysis examined the pattern of relations between variables to identify meaningful similarities and dissimilarities among the data. These analyses employed correlation coefficients to determine the direction and strength of associations among sets of variables. Differences in response patterns and outcomes across salient demographic and geographic variables were explored with commonly used statistic tools (Chi-square tests, t-tests, ANOVA statistics, and multiple regression analyses).

THRESHOLD SENSITIVITY OR "TASTE TEST" METHOD

Since this study focused essentially on the perception of tap water chlorinous flavors, a number of surveyed customers were recruited to undergo taste tests in addition to completing the survey. The description of how these tests were developed, implemented, and evaluated is detailed below.

Threshold Sensitivity to Chlorinous Flavors

Threshold sensitivity tests (referred below as the "Taste Test", for simplicity) were used to assess the average customer sensitivity to chlorinous flavors. This involved several steps: 1) Recruit approximately 30-40 average water utility customers; 2) Prepare the test materials; 3) Conduct the taste test and survey. Each of these steps is described in detail below.

Participant Recruitment

Average consumers from the participating water utility's service area were recruited via random digit dialing (except Midwest 2, which used utility personnel for the taste test due to recruiting difficulties). Participants were prescreened for disqualifying characteristics which might influence their sensitivity to chlorinous flavors (e.g., were pregnant or had a cold), or which might bias their opinions (work at a water utility). Respondents were asked to refrain from smoking, eating or drinking flavored beverages one hour before the test, and to refrain from wearing aftershave or perfume to the test.

Recruits were assigned specific test times. Time slots at noon, 5:00/5:30 p.m., and 6:30/7:00 p.m. were offered on two consecutive weekdays. To facilitate taste test management, a maximum of ten persons was assigned to each slot.

Material Preparation

ASTM Method E679-91, 1997, "Standard Practice for Determination of Odor and Taste Thresholds By a Forced-Choice Ascending Concentration Series Method of Limits," referred to here as the triangle test, was used to measure the participants' level of sensitivity to free chlorine or monochloramine in drinking water. The method can be summarized as follows: participants taste seven sets of three samples, each set consisting of two blanks and one chlorinated sample. In each set, the participants choose the sample that contains the flavor-causing substance (chlorine in this case). If no difference can be perceived, the participant must guess. The chlorine concentration increases from one set to the next. The best-estimate sensitivity limit for each panelist is defined as the geometric mean of that concentration at which the last miss occurred and the next (adjacent) higher concentration.

Glassware Preparation

All glassware preparation followed standard laboratory procedures and started with cleaning using a detergent and thorough rinsing. The chlorine demand of the 250-mL and 1-L amber glass bottles was then removed by filling the bottles with a chlorine solution of ~100 mg/L Cl_2 for at least 1 hour, followed by at least three rinses with tap water and then de-ionized (DI) water. The glassware was air-dried prior to use.

Collection of Water

To prepare the samples, finished water was collected at the plant effluent (after addition of chlorine) in NSF-certified 19-L (5 gallon) polycarbonate carboys (STC Plastics, Inc., Chino, California). In all cases, the drinking water collected met all federal and state requirements (including concentration x time [CT] criteria). Water quality data was also collected including the free and total chlorine residuals, pH, temperature, turbidity, conductivity, hardness, TDS, and alkalinity.

De-Chlorination

The free and total chlorine residuals of the collected water were measured and de-chlorination was performed based on the total chlorine level using sodium sulfite (Na_2SO_3, ACS grade; Fisher Scientific, Pittsburgh, Pennsylvania). Stoichiometric doses were used (1 mole of Na_2SO_3 to de-chlorinate 1 mole of Cl_2). De-chlorination was performed directly in the 19-L carboys, and the target residual was <0.10 mg/L Cl_2 for both free and total chlorine. Enough Na_2SO_3 solution was added to de-chlorinate the water without leaving any residual that could otherwise compromise the flavors during subsequent re-chlorination. The de-chlorinated water was used for blanks as well as to prepare the chlorinated samples for the triangle test.

Re-Chlorination/Re-Chloramination

The same type of disinfectant used at the facility where the taste tests were conducted was used for re-chlorination (i.e., utilities using free chlorine to maintain residual were tested with free chlorine and the same was done with monochloramines).

A free chlorine solution (100 mg/L Cl_2) was prepared from sodium hypochlorite (NaOCl, 6%; VWR Scientific Products, Willard, Ohio), and its pH was adjusted to match the local water pH so the speciation of the chlorine compounds in the test samples was commensurate with the tap water.

For rechloramination, a 50 mg NH_3-N/L (West Coast site, ratio Cl_2:NH_3 of 3:1) or 30 mg NH_3-N/L (Southeast 2 site, ratio Cl_2:NH_3 of 5:1) and 40 mg NH_3-N/L (Midwest 2 site, ratio Cl_2:NH_3 of 4:1) was prepared. The Cl_2: N-NH_3 mass ratios used were similar to those used at each plant. Monochloramine solutions were prepared by adding the solution of ammonium dichloride (NH_4Cl) to the free chlorine solution at a pH above 8.5 (to avoid undesirable formation of di- and tri-chloramines [Beck et al. 1986]).

The targeted chlorine doses were selected according to ASTM method E679-91, which specifies geometric increases in concentration (two adjacent concentrations are separated by a constant factor, Table 3.8). The limits of sensitivity reported in the literature were used to select the "X" value: 0.25 mg/L for free chlorine and 0.50 mg/L for monochloramine (Krasner and Barrett 1984). Seven concentrations were used to maximize precision, while limiting participants' saturation during tasting, which occurs when too many samples are tested. In the case of monochloramine, the test values were raised to a maximum concentration (#7) near the maximum residual disinfectant level (MRDL) of 4.0 mg/L Cl_2.

All free and total chlorine residuals were measured in duplicate, or triplicate if duplicate results showed differences greater than 0.3 mg/L. After the first monochloramine test (Southeast 2 Market), the sensitivity limit and multiplier were changed, because the literature-based values were not yielding good results - the sensitivity threshold was higher than reported. Therefore, the sensitivity limit was increased to 0.750 mg/L and 1.5 was chosen as the multiplier in order to give a range of concentrations that would not exceed 4 mg/L (the regulated MRDL permitted on the distribution system).

Table 3.8
Target concentrations

	Free chlorine			Monochloramines			
#	Target concentrations (mg/L)	Multiplier	#	Target concentrations* (mg/L)		Multiplier*	
1	0.167	X/1.5	1	0	0.333		$X/1.5^2$
2	0.250	X	2	0.154	0.500	$X/1.8^2$	X/1.5
3	0.3750	1.5 X	3	0.278	0.750	X/1.8	X
4	0.563	1.5^2 X	4	0.500	1.125	X	X/1.5
5	0.844	1.5^3 X	5	0.9	1.688	1.8X	1.5^2X
6	1.266	1.5^4 X	6	1.620	2.531	1.8^2X	1.5^3X
7	1.898	1.5^5 X	7	2.916	3.792	1.8^3X	1.5^4X

* The multiplier of 1.8 and X = 0.5 were used for Southeast 2 market, while a multiplier of 1.5 and X = 0.75 were used for Midwest 2 and West coast markets.

Sample Preparation

A 1-liter solution of the highest concentration to be tested was prepared by adding the prepared chlorine or chloramine solution to the de-chlorinated water (from each utility market) to achieve the highest concentration (#7). All remaining concentrations were obtained through dilution of this solution with de-chlorinated water. This procedure eliminated any potential artifacts that could have occurred due to the addition of effluent volumes of chlorinous solutions.

One-liter chlorine free amber bottles with Teflon™ caps were used to prepare all solutions. The chlorinated samples were prepared as late as possible (no more than 2-3 hours prior to the taste test) and kept at room temperature in darkness. The pH was adjusted to meet the finished water value (around 9 for monochloramine solutions), and test temperature was recorded for each test.

The free and total chlorine concentrations were measured using the Hach pocket colorimeter chlorine test kits (Hach Company, Loveland, Colorado) according to USEPA Method 330.5. Separate 10-mL cuvettes were used for free and total chlorine, and residuals measured in duplicate, or triplicate if the difference between the duplicates was greater than 0.30 mg/L Cl_2. When possible, amperometric titration was also used to verify the results obtained with the colorimetric method.

Taste Tests

Participants were escorted to the testing room and provided with taste test instructions, an answer sheet, and survey questionnaire (examples of these documents are located in Appendix C). The room was selected to provide optimal test conditions (e.g., quiet, free of odors, such as coffee and cleaning products, clean, uniform lighting, regular temperature and easy access). Interaction between panelists was minimized.

During the taste test, each participant evaluated seven sets of three samples (one containing chlorinated water, and two containing de-chlorinated water), as described in the ASTM method E679-91. Because several panelists were being tested simultaneously, a matrix was created to eliminate the possibility for a participant to be influenced by another participant's answers (an extract of the answer table is provided in Table 3.9). About 30 mL of blanks (de-chlorinated water) or samples (chlorinated water) were poured in 3 ½ oz (104 mL) colored, plastic cups (Solo Cup Company, Urbana, Illinois), and capped with aluminum foil. The blanks were poured first (prior to the beginning of the test), and the samples containing chlorine were poured at the very last minute (less than 3 minutes before tasting) to limit chlorine evaporation.

"Solo" cups were selected because they were tasteless and odorless. Because the shiny side of the aluminum foil contains chemicals that may impart taste to the water, all foil covers were placed "dull" side down on the cups. A 2-minute recovery time was recommended between each set to limit sensory fatigue. The participants also answered the survey questionnaire to allow the development of correlations between their perception of tap water and their objective sensitivity to chlorine.

Table 3.9
Extract of answer matrix

Sample number	Panelist #1			Panelist #2			Panelist #3			Panelist #4		
	A	B	C	A	B	C	A	B	C	A	B	C
1	+	O	O	O	+	O	O	O	+	O	+	O
2	O	+	O	O	O	+	O	+	O	+	O	O
3	O	O	+	O	+	O	O	+	O	O	O	+
4	+	O	O	O	+	O	+	O	O	O	+	O
5	O	O	+	+	O	O	O	O	+	+	O	O
6	O	O	+	O	O	+	+	O	O	O	O	+
7	+	O	O	O	+	O	O	O	+	O	+	O

\+ Chlorinated sample
O De-chlorinated water (blank)

Data Analysis

Immediately before and after the taste test, the water temperature, pH, free and total chlorine residuals of each sample were measured. To determine the sensitivity limits, the concentrations considered are the averages between the levels measured before and after the test. The participants' answers were analyzed and their threshold sensitivity limits determined according to the ASTM method E679-91 best-estimate criterion (the geometric mean of the last missed concentration and the next higher level).

CHAPTER 4
NATIONAL AND MARKET OPINION SURVEYS AND TASTE TESTS - RESULTS

INTRODUCTION

The results of the national and regional market opinion surveys and the results of the 7-market consumer threshold sensitivity (to free or combined chlorine) tests are summarized in this chapter. The full telephone survey results are available in Appendix F on the CD-ROM that accompanies this report.

Among other aspects, consumer perception of tap water quality, level of education, and awareness and taste test results are shown and discussed. More specifically, for each market, this chapter summarizes physical data, describes how surveyed consumers perceived their tap water, assesses what percentage of the population searched for tap water alternatives and what caused that transition, and relates that to the results of taste tests.

OBJECTIVE

There were three major objectives in the analysis: 1) Define consumer perceptions of tap water quality, particularly with respect to chlorinous flavor; 2) Determine which tap water parameters mostly affected and impacted consumer perceptions of tap water quality and safety; 3) Characterize consumer perceptions and sensitivity to chlorinous flavors.

METHODOLOGY

The methodology of the telephone survey and consumer taste test is described in detail in Chapter 3. The full survey instrument is included in Appendix A.

RESULTS

National Market Survey

Consumer Perception of Water Quality

The overall results of the national survey are summarized in Table 4.1. Consumer descriptions of tap water taste and smell are illustrated in Figures 4.1 and 4.2. Although 66% of consumers were satisfied or very satisfied with their tap water taste, 54% identified a some sort of negative[1] taste in it; Despite the fact that 80% of consumers were satisfied or very satisfied with tap water smell, 16% detected some sort of negative smell in it.[1] The large majority of consumers perceived their tap water to be safe or somewhat safe (82%) and a somewhat smaller majority perceived it to be healthy or somewhat healthy (67.5%).

The most common "off" flavor (not "OK," "good," "fresh" or "nothing") reported was "chlorinous" (15.5% and 14.8% of respondents for taste and odor respectively). Consumers who perceived a chlorinous flavor were less satisfied than consumers who reported "none/nothing/fresh," but more satisfied than consumers who reported other undesirable flavors.

[1] Where negative is defined as responses other than "none," "nothing," "fresh," "OK," "good"...

Table 4.1
Consumer perception of water quality - National survey

	Very satisfied (%)	Satisfied (%)	Neither dissatisfied nor satisfied (%)	Dissatisfied (%)	Very dissatisfied (%)
Overall TW quality	41.1	31.3	2.0	13.0	12.6
Overall TW taste	37.6	28.1	1.2	15.7	17.4
Overall TW appearance	53.4	29.7	0.8	9.6	6.5
Overall TW smell	52.1	28.1	2.7	7.3	9.8
Description of TW safety	Safe	Somewhat safe	Neither unsafe nor safe	Somewhat unsafe	Unsafe
	64.7	17.3	3.3	9.0	5.7
Description of TW healthiness	Healthy	Somewhat healthy	Neither unhealthy nor healthy	Somewhat unhealthy	Unhealthy
	49.6	17.9	11.6	15.1	5.8

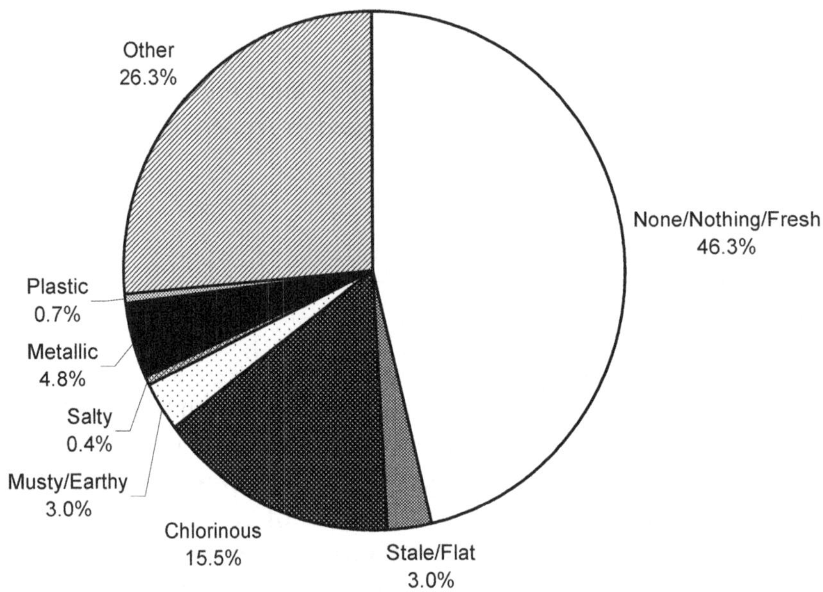

Figure 4.1 Range of tastes reported by consumers - National survey

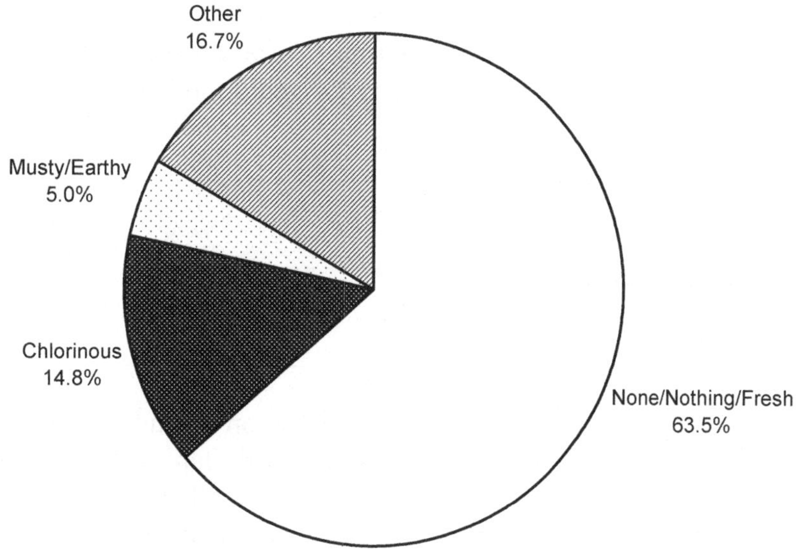

Figure 4.2 Range of smells reported by consumers - National survey

Consumer confidence in tap water quality compliance with drinking water regulations is comparable to perceptions of tap water safety (Table 4.2). Not surprisingly, consumers who were confident of tap water safety were more likely to be confident that it met drinking water regulatory requirements.

Table 4.2
Consumer responses to the question "How confident are you that the tap water you receive in your household meets or exceeds state and federal requirements?" - National survey

Response	% of respondents
Very confident	24.8
Confident	46.1
Neither not confident nor confident	12.1
Not confident	9.9
Not very confident	7.1

Three other statements about tap water safety and chlorinous flavor were asked (Table 4.3). While consumers in general believed that tap water is safe and that the scientific and medical community have a good understanding of tap water-related health risks, only about a third of the population was in agreement with the statement that it is safe to drink chlorinous-flavored water.

Table 4.3
Consumer responses to the statements listed below - National survey

	"In general, I believe it is safe to drink tap water." (%)	"I believe it is safe to drink water that has a chlorinous taste or odor." (%)	"Scientists and medical experts have a good understanding of any health risks from drinking tap water." (%)
Strongly agree	21.2	4.1	12.9
Agree	51.3	25.8	53.1
Neither disagree nor agree	11.7	12.3	17.9
Disagree	10.1	40.4	12.8
Strongly disagree	5.7	17.4	3.2

These results indicate that consumer perception of undesirable tastes and odors does not necessarily result in dissatisfaction or distrust with tap water quality, and that consumer perception of chlorinous flavors is generally less unfavorable than perception of other off flavors, and consumers are generally trusting of the government and their water utility's ability to provide them with safe drinking water. However, there is still a significant portion of the population that is distrusting of the water industry's ability to provide safe drinking water.

Consumers were also asked whether they had contacted their local or municipal water provider with a complaint or question within the past year. Almost 90% of consumers surveyed indicated that they did not. Thus, the issues causing dissatisfaction with tap water quality and taste were generally not significant enough to motivate people to formally complain about them; or, consumers felt the utility could not/would not address their problem.

Consumer Education

Consumers were asked a series of questions to evaluate how informed they felt they were with respect to their tap water. Specifically, they were asked whether they received a "water quality report" (Consumer Confidence Report) within the last year, and if so, whether the report was useful. This phrasing was chosen to eliminate any potential bias from implying that they did receive one and so should remember (e.g., "Do you remember receiving your Water Quality Report this year?"). Since all utilities surveyed were in compliance with CCR regulations, it is assumed that all respondents were sent CCRs.

Only 30% of the survey respondents remembered receiving a CCR. This indicates that the mechanisms currently used to inform the consumers are insufficient and/or not captivating enough to the general public. Eighty percent of the consumers who remembered receiving the CCR found it helpful and were more positive with respect to tap water aesthetics and safety, so increasing consumer awareness should positively impact consumer satisfaction with tap water (Table 4.4).

Table 4.4
Consumer awareness of their CCR - National survey

	Yes (%)	No (%)
Received CCR in the last year	29.8	70.2
Report was useful	78.9	21.1

Target Market Surveys

As mentioned previously, approximately 200 people were surveyed in each of seven target markets. The results from these surveys are summarized and discussed below. A summary of treated water quality is presented in Tables 4.5 and 4.6. As expected from the wide distribution of target markets, water quality and consumer perception varied significantly.

In the following sections the results of each of the seven target markets is presented separately and then discussed at the end of the chapter.

Table 4.5
General information collected for each of the sampled markets

Market	Water source	Ownership	Utility size	T&O or safety issues
East Coast A*	SW	Private	Moderate	Summer 2000 - Boil order
East Coast B*	SW	Private	Moderate	T&O in the summer (earthy musty odors)
	GW			Sulfur related odors
Southeast 1	GW (aquifer recharged with wastewater)	Public	Large	High TOC for GW (2.5 mg/L) turbidity spikes (highest value recorded - 20 NTU)
Midwest 1	SW Lake Michigan	Public	Large	Spring - T&O 2-methyl isoborneol (MIB)/ Geosmin
West	GW - 85% GWUDI - 15%	Private	Moderate	
Southeast 2	GW	Public	Moderate	
Midwest 2		Public	Moderate	Aesthetics are worse in hot weather
West Coast	SW Mokelumne River (snowmelt)	Public	Large	1998-Switched from free to carbon chlorine 1999-T&O decreased

* Two East Coast water utilities' service areas were combined to form the "East Coast" market.
SW: Surface water
GW: Ground water
GWUDI: Groundwater under the direct influence of surface water

Table 4.6
Treated water quality data for participating utilities

Market	Temperature (°C)		Hardness (mg/L as $CaCO_3$)		TDS (mg/L)		Type of chlorine residual	Chlorine residual* (mg/L as Cl_2)	
	Avg.	Range	Avg.	Range	Avg.	Range		Avg.	Range
East Coast	13	1-34	70	48-92	138	116-159	Free	0.9	0.4-1.5
Southeast 1	26	25-28	134	129-140	172	157-185	Free	6.6	1.4-13.1
Midwest1	11	2-23	138	134-144	158	156-164	Free	0.7	0.6-0.8
West	19	14-29	104	15-241	206	36-551	Free	1.0	0.7-1.0
Southeast 2	23	11-24	182	-	480	-	Combined	4.0	0.6-5.2-
Midwest 2	13	2-25	131	78-191	383	121-655	Combined	2.3	1.5-2.8
West Coast	14	10-18	35	10-130	71	33-241	Combined	2.2	1.9-2.5

* Measured at the plant effluent (not the first customer's tap).

Free Chlorine Markets

East Coast Market

Consumer perception of water quality. A summary of consumers' responses to questions about their household tap water quality, taste, appearance, smell, safety, and healthiness is provided in Table 4.7. Consumers were also asked to describe the taste and smell of their household tap water; responses are illustrated in Figures 4.3 and 4.4.

Table 4.7
Consumer perception of water quality - East Coast market survey

	Very satisfied	Satisfied	Neither dissatisfied nor satisfied	Dissatisfied	Very dissatisfied
Overall TW quality	41.1	39.4	2.5	11.2	5.8
Overall TW taste	38.8	31.8	1.7	17.0	10.7
Overall TW appearance	71.2	23.5	1.9	2.6	0.8
Overall TW smell	61.3	23.2	1.5	8.5	5.5
Description of TW safety	Safe	Somewhat safe	Neither unsafe nor safe	Somewhat unsafe	Unsafe
	64.9	20.8	4.8	5.7	3.8
Description of TW healthiness	Healthy	Somewhat healthy	Neither unhealthy nor healthy	Somewhat unhealthy	Unhealthy
	48.5	25.8	8.2	13.1	4.4

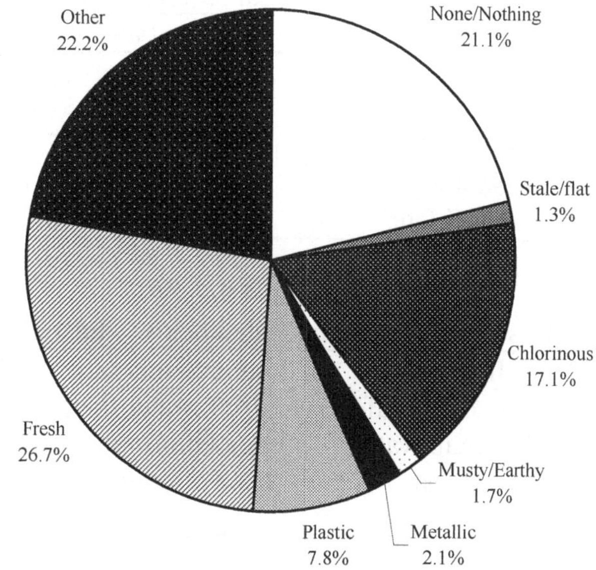

Figure 4.3 Consumer perception of tap water taste - East Coast market survey

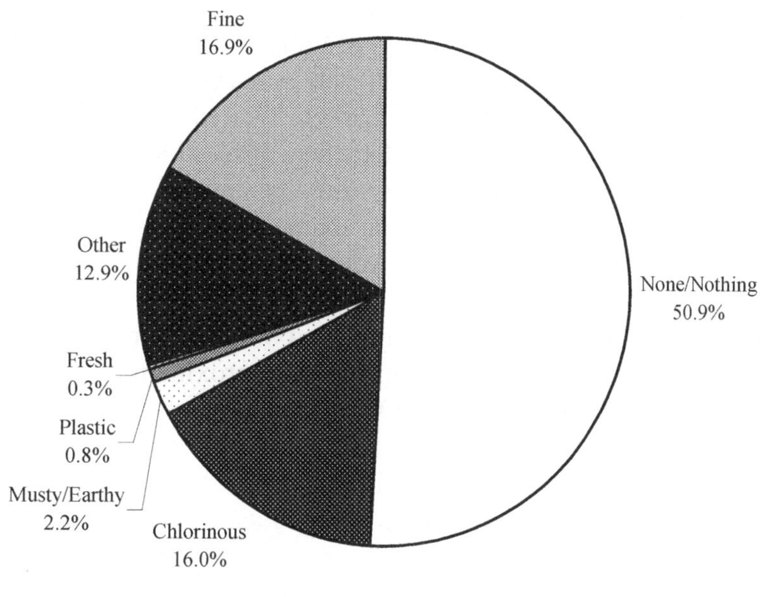

Figure 4.4 Consumer perception of tap water smell - East Coast market survey

Perception of tap water safety. The majority of respondents were confident their tap water met drinking water regulatory requirements (Table 4.8), though almost a third were not. The majority perceived their tap water to be safe to drink and scientists' and physicians' understanding of tap water-related health risks to be good, though consumers were significantly split on their perception of the safety of chlorinous flavors (Table 4.9).

Table 4.8

Consumer responses to the question "How confident are you that the tap water you receive in your household meets or exceeds state and federal requirements?" - East Coast market survey

Response	% of respondents
Very confident	18.8
Confident	52.6
Neither not confident nor confident	17.1
Not confident	5.4
Not very confident	6.1

Table 4.9

Consumer responses to the statements listed below - East Coast market survey

	"In general, I believe it is safe to drink tap water." (%)	"I believe it is safe to drink water that has a chlorinous taste or odor." (%)	"Scientists and medical experts have a good understanding of any health risks from drinking tap water." (%)
Strongly agree	21.4	3.8	11.7
Agree	50.6	33.2	50.2
Neither disagree nor agree	13.5	9.1	22.1
Disagree	10.3	35.7	13.4
Strongly disagree	4.4	18.2	2.6

Primary drinking water source. Consumers were asked whether they primarily drink tap water or a tap water alternative (TWA - filtered water or bottled water) while at home. Forty-four percent of East Coast market consumers surveyed indicated that they drink primarily tap water in their household leaving over half of consumers who primarily drink some form of TWA. Among the filtered water drinkers in the East Coast market, 29% said they drank filtered water mainly for taste reasons, while 29 and 25% indicated they drink filtered water primarily for overall quality and safety reasons, respectively. Thirty-three percent of the bottled water drinkers in the East Coast market indicated that taste was the main reason they chose to drink bottled water, while 23 and 22% said they drink bottled water primarily due to overall quality and health factors, respectively.

Consumer education. Consumers were asked whether they had received a CCR within the last year. Table 4.10 summarizes their answers. Approximately one-third of the population recalled receiving their water quality report. Of those who received it, most found it useful.

Table 4.10

Consumer awareness of their CCR - East Coast market survey

	Yes	No
Received CCR in the last year	35.4	64.6
Report was useful	78.6	21.4

Threshold sensitivity to free chlorine. A total of 41 panelists were surveyed in the East Coast market. The results are summarized in Table 4.11. The average measured consumer sensitivity to free chlorine was 0.77 mg/L Cl_2 . As would be expected with a randomly-selected human population, there was a large variation in sensitivity (Figure 4.5). Sixty-three percent of the panelists had a measured sensitivity level greater than 0.90 mg/L Cl_2 (the average residual concentration at the first customer's tap) the majority of which were significantly higher (> 1.0 mg/L Cl_2) than the effluent residual.

Table 4.11
Measured threshold sensitivities of tap water (TW), bottled water (BW), filtered water (FW), and the group as a whole (all) to free chlorine - East Coast market panelists

User group	# panelists	Log avg. threshold	Avg. threshold	Log S.D.	Max	Min	Median
TW	20	-0.21	0.62	0.48	2.31	0.12	0.80
BW	8	-0.04	0.91	0.25	2.25	0.40	1.00
FW	13	0.00	0.99	0.36	2.28	0.19	1.39
All	41	-0.11	0.77	0.41	2.31	0.12	1.00

S.D. Standard deviation

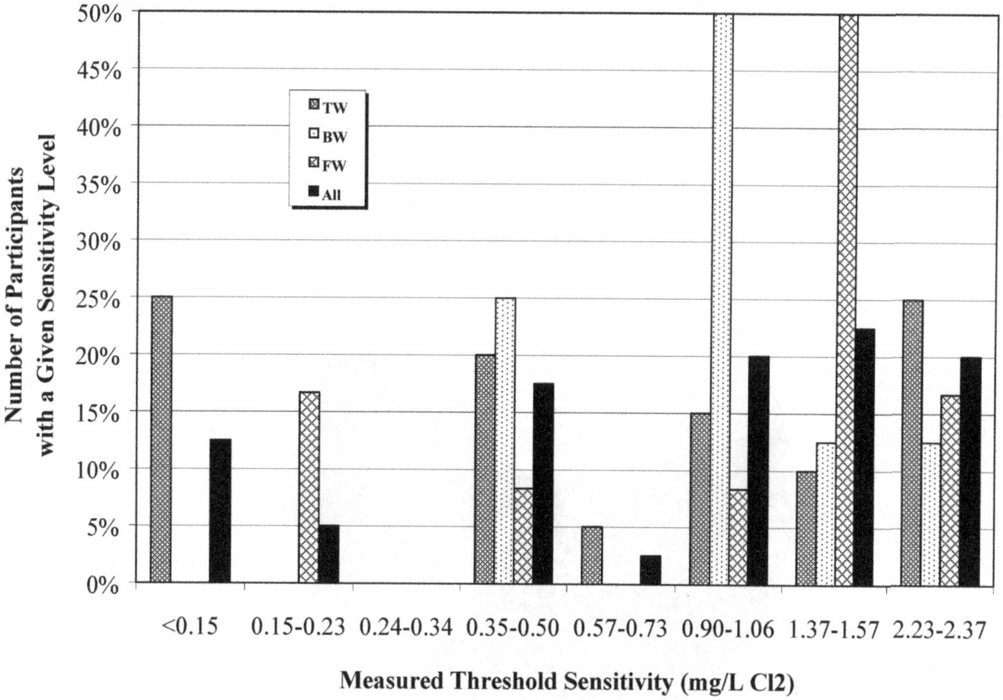

Figure 4.5 Distribution of measured threshold sensitivities to free chlorine - East Coast market panelists

Southeast 1 Market

Consumer perception of water quality. A summary of consumers' responses to questions about their household tap water quality, taste, appearance, smell, safety, and healthiness is provided in Table 4.12. Consumers were also asked to describe the taste and smell of their household tap water; responses are illustrated in Figures 4.6 and 4.7.

Table 4.12
Consumer perception of water quality - Southeast 1 market survey

	Very satisfied (%)	Satisfied (%)	Neither dissatisfied nor satisfied (%)	Dissatisfied (%)	Very dissatisfied (%)
Overall TW quality	34.9	30.8	1.3	20.7	12.3
Overall TW taste	32.6	32.0	1.6	17.1	16.7
Overall TW appearance	57.4	29.9	0.3	4.1	8.3
Overall TW smell	49.6	28.0	1.6	10.8	10.0
Description of TW safety	Safe	Somewhat safe	Neither unsafe nor safe	Somewhat unsafe	Unsafe
	52.1	27.4	4.2	11.7	4.6
Description of TW healthiness	Healthy	Somewhat healthy	Neither unhealthy nor healthy	Somewhat unhealthy	Unhealthy
	39.9	22.6	16.7	13.2	7.6

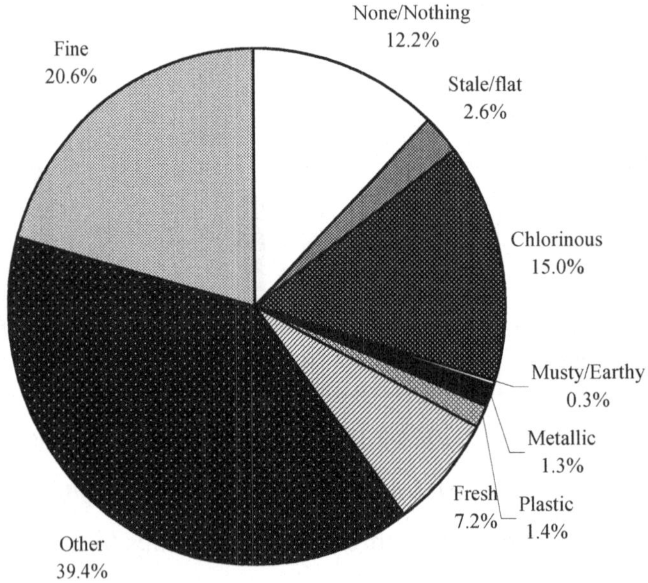

Figure 4.6 Consumer perception of tap water taste - Southeast 1 market survey

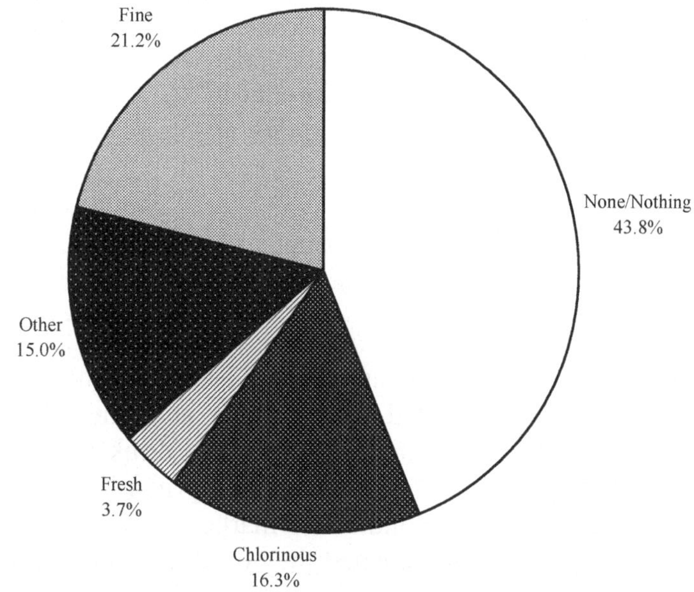

Figure 4.7 Consumer perception of tap water smell - Southeast 1 market survey

Perception of tap water safety. The great majority of respondents were confident their tap water met drinking water regulatory requirements (Table 4.13), though one quarter were not. The majority perceived their tap water to be safe to drink and scientists' and physicians' understanding of tap water-related health risks to be good, though consumers were significantly split on their perception of the safety of chlorinous flavors (Table 4.14).

Table 4.13
Consumer responses to the question "How confident are you that the tap water you receive in your household meets or exceeds state and federal requirements?" - Southeast 1 market survey

Response	% of respondents
Very confident	22.0
Confident	53.1
Neither not confident nor confident	10.0
Not confident	8.9
Not very confident	6.0

Table 4.14

Consumer responses to the statements listed below - Southeast 1 market survey

	"In general, I believe it is safe to drink tap water." (%)	"I believe it is safe to drink water that has a chlorinous taste or odor." (%)	"Scientists and medical experts have a good understanding of any health risks from drinking tap water." (%)
Strongly agree	15.2	8.1	15.3
Agree	52.7	26.3	54
Neither disagree nor agree	13.5	11.1	19.3
Disagree	10.8	35.0	10.8
Strongly disagree	7.8	19.5	0.7

Primary drinking water source. Consumers were asked whether they primarily drink tap water or a TWA (filtered water or bottled water) while at home. Twenty-seven percent of Southeast 1 market consumers surveyed indicated that they drink primarily tap water in their household, leaving nearly three-quarters of consumers who primarily drink some form of TWA. Among the filtered water drinkers in the Southeast 1 market, 35% said they drink filtered water mainly for taste, while 31, 17, and 17% indicated they drink filtered water primarily for health, safety, and overall quality reasons, respectively. Twenty-eight percent of the bottled water drinkers in the Southeast 1 market indicated that taste was the main reason they chose to drink bottled water, while 20, 18, and 16% said they drink bottled water primarily due to overall quality, safety, and health factors, respectively.

Consumer education. Consumers were asked whether they had received a CCR within the last year. Table 4.15 summarizes their answers. Approximately one quarter of the population recalled receiving their water quality report. Of those who received it, most found it useful.

Table 4.15

Consumer awareness of their CCR - Southeast 1 market survey

	Yes	No
Received CCR in the last year	22.5	77.5
Report was useful	93.8	6.2

Threshold sensitivity to free chlorine. A total of 40 panelists were surveyed in the Southeast 1 market. The results are summarized in Table 4.16. The average measured consumer sensitivity to free chlorine was 1.1 mg/L Cl_2 (Figure 4.8). Forty-five percent of the panelists had a measured sensitivity level greater than 1.0 mg/L Cl_2, the average residual concentration at the first customer's tap.

Table 4.16

Measured threshold sensitivities of TW, BW, FW, and the group as a whole (all) to free chlorine - Southeast 1 market panelists

User group	# of panelists	Log avg. threshold	Avg. threshold (mg/L cl_2)	Log S.D.	Max	Min	Median
TW	16	0.11	1.27	0.34	2.34	0.19	1.89
BW	12	-0.08	0.83	0.42	2.37	0.13	0.81
FW	12	0.07	1.19	0.30	2.37	0.38	1.18
All	40	0.04	1.10	0.36	2.37	0.13	1.41

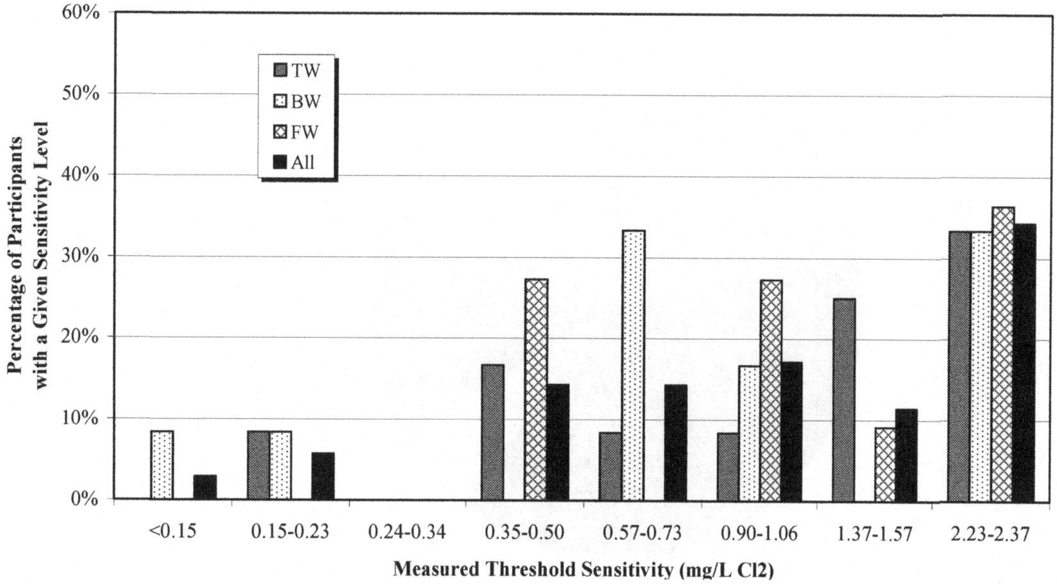

Figure 4.8 Distribution of measured threshold sensitivities to free chlorine - Southeast 1 market panelists

Midwest 1 Market

Consumer perception of water quality. A summary of consumers' responses to questions about their household tap water quality, taste, appearance, smell, safety, and healthiness is provided in Table 4.17. Consumers were also asked to describe the taste and smell of their household tap water; responses are illustrated in Figures 4.9 and 4.10.

Table 4.17
Consumer perception of water quality - Midwest 1 market survey

	Very satisfied	Satisfied	Neither dissatisfied nor satisfied	Dissatisfied	Very dissatisfied
Overall TW quality	48.7	33.0	1.6	10.5	6.2
Overall TW taste	43.5	34.0	0.0	17.7	4.8
Overall TW appearance	63.5	26.7	0.0	6.9	2.9
Overall TW smell	64.8	24.0	2.5	5.4	3.3
Description of TW safety	Safe	Somewhat safe	Neither unsafe nor safe	Somewhat unsafe	Unsafe
	61.8	22.9	5.7	7.9	1.7
Description of TW healthiness	Healthy	Somewhat healthy	Neither unhealthy nor healthy	Somewhat unhealthy	Unhealthy
	53.4	19.7	16.5	9.0	1.4

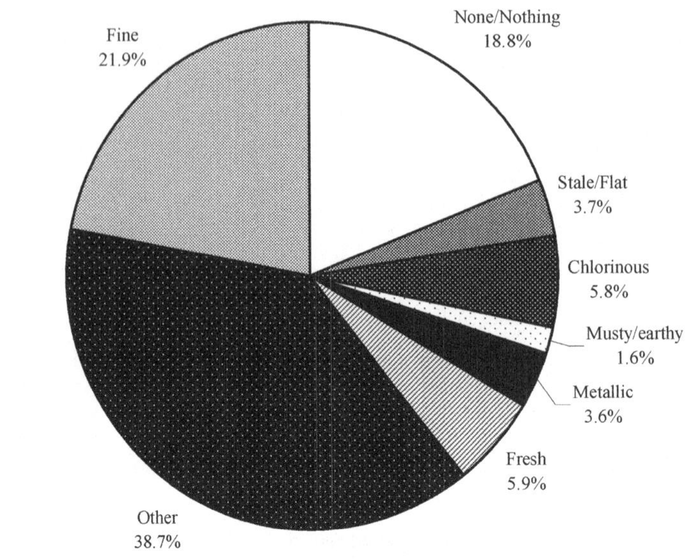

Figure 4.9 Consumer perception of tap water taste - Midwest 1 market survey

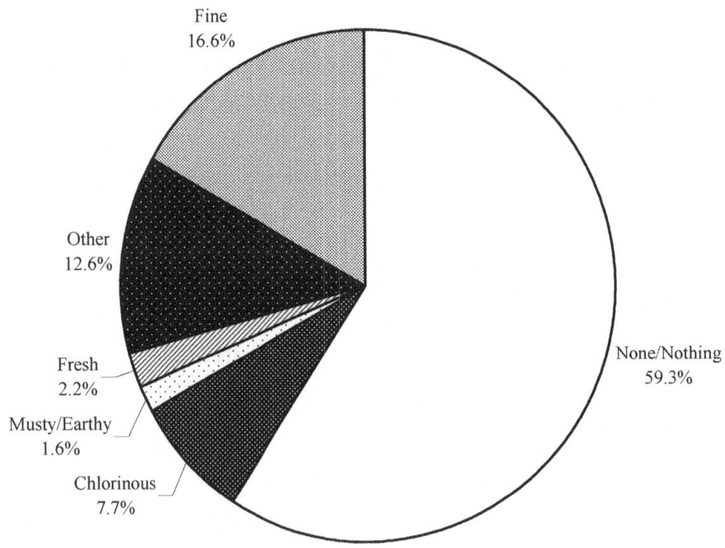

Figure 4.10 Consumer perception of tap water smell - Midwest 1 market survey

Perception of tap water safety. The great majority of respondents were confident their tap water met drinking water regulatory requirements (Table 4.18), though almost one quarter were not. The majority perceived their tap water to be safe to drink and scientists' and physicians' understanding of tap water-related health risks to be good, though consumers were significantly split on their perception of the safety of chlorinous flavors (Table 4.19).

Table 4.18
Consumer responses to the question "How confident are you that the tap water you receive in your household meets or exceeds state and federal requirements?" - Midwest 1 market survey

Response	% of respondents
Very confident	22.1
Confident	55.2
Neither not confident nor confident	13.7
Not confident	5.1
Not very confident	3.9

Table 4.19
Consumer responses to the statements listed below - Midwest 1 market survey

	"In general, I believe it is safe to drink tap water." (%)	"I believe it is safe to drink water that has a chlorinous taste or odor." (%)	"Scientists and medical experts have a good understanding of any health risks from drinking tap water." (%)
Strongly agree	24.5	3.7	15.6
Agree	49.0	32.2	59.6
Neither disagree nor agree	10.8	16.3	13.2
Disagree	10.2	35.0	8.6
Strongly disagree	5.6	12.8	3.0

Primary drinking water source. Consumers were asked whether they primarily drink tap water or a tap water alternative (filtered water or bottled water) while at home. Fifty-five percent of Midwest 1 market consumers surveyed indicated that they drink primarily tap water in their household, leaving 45% of consumers who primarily drink some form of TWA. Among the filtered water drinkers in the Midwest 1 market, 43% said they drink filtered water mainly for overall quality, while 34 and 13% indicated they drink filtered water primarily for safety and health reasons, respectively. Twenty-two percent of the bottled water drinkers in the Midwest 1 market indicated that taste was the main reason they chose to drink bottled water, while 19, 17, and 12% said they drink bottled water primarily due to health, overall quality, and safety factors, respectively.

Consumer education. Consumers were asked whether they had received a CCR within the last year. Table 4.20 summarizes their answers. Approximately one-third of the population recalled receiving their water quality report. Of those who received it, most found it useful.

Table 4.20
Consumer awareness of their CCR - Midwest 1 market survey

	Yes	No
Received CCR in the last year	33.8	66.2
Report was useful	70.2	29.8

Threshold sensitivity to free chlorine. A total of 30 panelists were surveyed in the Midwest 1 market. The results are summarized in Table 4.21. The average measured consumer sensitivity to free chlorine was 1.02 mg/L Cl_2 (Figure 4.11). Seventy-nine percent of the panelists had a measured sensitivity level greater than 0.70 mg/L Cl_2 (the residual concentration at the first customer's tap), the majority of which were significantly higher (> 1.3 mg/L Cl_2) than the effluent residual.

Table 4.21
Measured threshold sensitivities of TW, BW, FW, and the group as a whole (all) to free chlorine - Midwest 1 market panelists

User group	# of panelists	Log avg. threshold	Avg. threshold (mg/L cl_2)	Log S.D.	Max	Min	Median
TW	14	0.04	1.09	0.50	2.36	0.09	2.23
BW	8	0.08	1.21	0.17	2.28	0.72	1.26
FW	8	-0.12	0.76	0.60	2.36	0.09	1.27
All	30	0.01	1.02	0.46	2.36	0.09	1.49

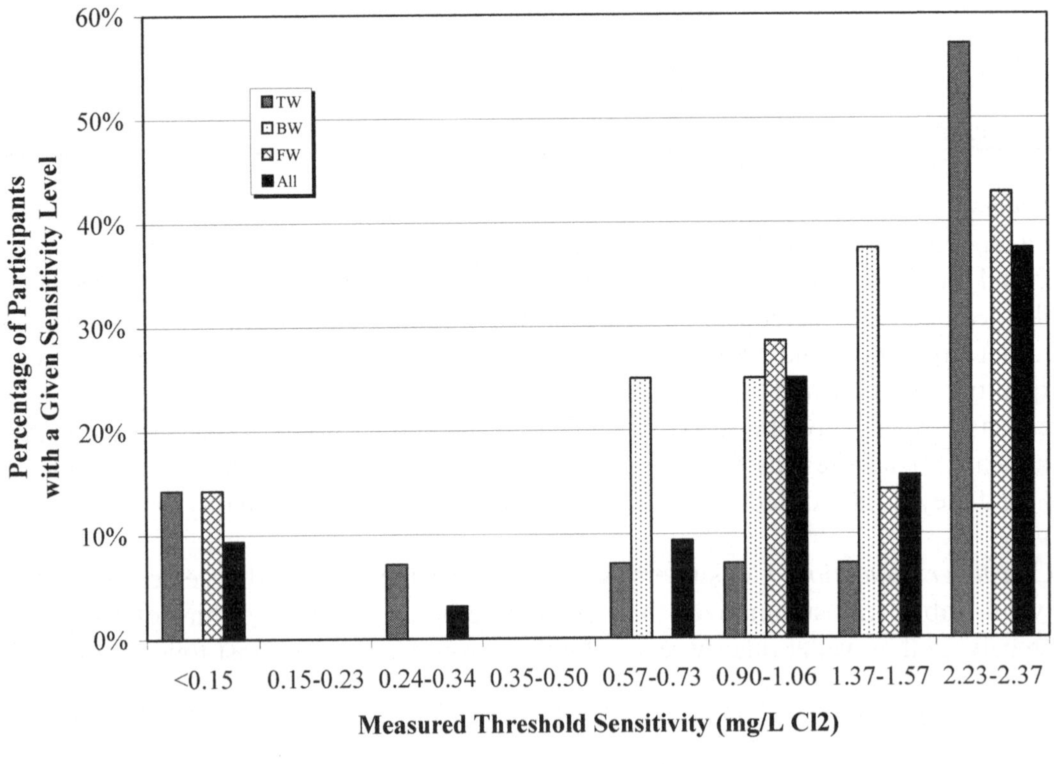

Figure 4.11 Distribution of measured threshold sensitivities to free chlorine - Midwest 1 market panelists

West Market

Consumer perception of water quality. A summary of consumers' responses to questions about their household tap water quality, taste, appearance, smell, safety, and healthiness is provided in Table 4.22. Consumers were also asked to describe the taste and smell of their household tap water; responses are illustrated in Figures 4.12 and 4.13.

Table 4.22
Consumer perception of water quality - West 1 market survey

	Very satisfied	Satisfied	Neither dissatisfied nor satisfied	Dissatisfied	Very dissatisfied
Overall TW quality	41.0	32.8	2.5	12.5	11.2
Overall TW taste	38.7	28.1	1.0	13.6	18.6
Overall TW appearance	58.7	29.7	1.5	7.0	3.1
Overall TW smell	52.2	27.3	4.1	8.9	7.5
Description of TW safety	Safe	Somewhat safe	Neither unsafe nor safe	Somewhat unsafe	Unsafe
	64.8	18.8	8.7	3.2	4.5
Description of TW healthiness	Healthy	Somewhat healthy	Neither unhealthy nor healthy	Somewhat unhealthy	Unhealthy
	54.3	22.5	7.7	10.5	5.0

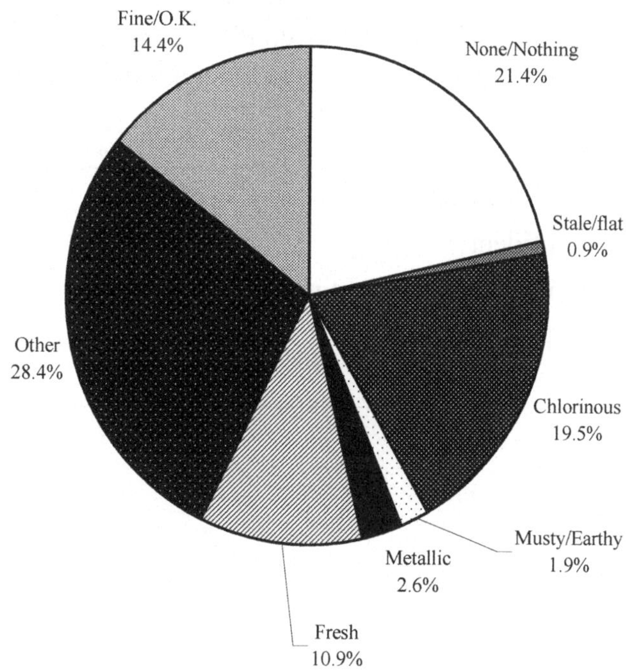

Figure 4.12 Consumer perception of tap water taste - West market survey

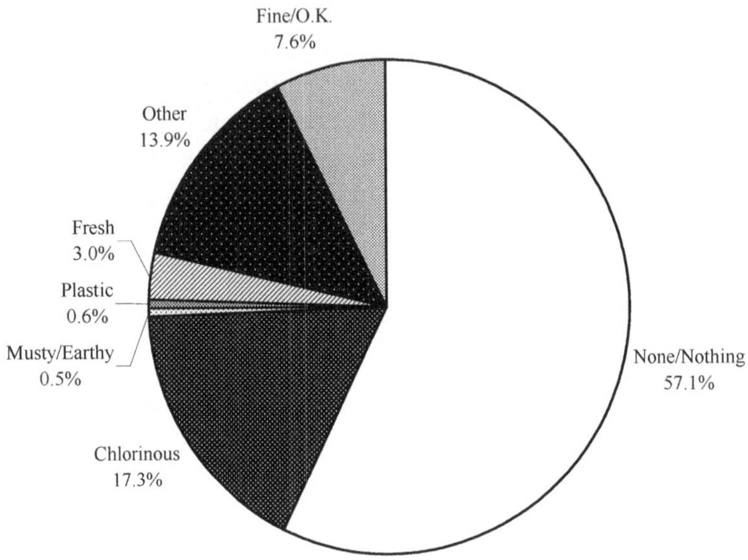

Figure 4.13 Consumer perception of tap water smell - West market survey

Perception of tap water safety. The great majority of respondents were confident their tap water met drinking water regulatory requirements (Table 4.23), though 20% were not. The majority perceived their tap water to be safe to drink and scientists' and physicians' understanding of tap water-related health risks to be good, though consumers were significantly split on their perception of the safety of chlorinous flavors (Table 4.24).

Table 4.23
Consumer responses to the question "How confident are you that the tap water you receive in your household meets or exceeds state and federal requirements?" - West market survey

Response	% of respondents
Very confident	33.0
Confident	45.6
Neither not confident nor confident	10.0
Not confident	2.9
Not very confident	8.4

Table 4.24
Consumer responses to the statements listed below - West market survey

	"In general, I believe it is safe to drink tap water." (%)	"I believe it is safe to drink water that has a chlorinous taste or odor." (%)	"Scientists and medical experts have a good understanding of any health risks from drinking tap water." (%)
Strongly agree	25.4	8.1	18.0
Agree	57.4	45.3	62.8
Neither disagree nor agree	4.1	6.1	10.3
Disagree	7.0	26.1	4.0
Strongly disagree	6.1	14.4	4.9

Primary drinking water source. Consumers were asked whether they primarily drink tap water or a tap water alternative (filtered water or bottled water) while at home. Fifty-seven percent of West market consumers surveyed indicated that they drink primarily tap water in their household, leaving 43% of consumers who primarily drink some form of TWA. Among the filtered water drinkers in the West market, 34% said they drink filtered water mainly for health reasons, while 30, 11, and 11% indicated they drink filtered water primarily for overall quality, taste, and safety reasons, respectively. Forty-two percent of the bottled water drinkers in the West market indicated that taste was the main reason they chose to drink bottled water, while 16, 16, and 11% said they drink bottled water primarily for health, overall quality, and safety reasons, respectively.

Consumer education. Consumers were asked whether they had received a CCR within the last year. Table 4.25 summarizes their answers. Approximately one-third of the population recalled receiving their water quality report. Of those who received it, most found it useful.

Table 4.25
Consumer awareness of their CCR - West market survey

	Yes (%)	No (%)
Received CCR in the last year	30.4	69.6
Report was useful	73.2	26.8

Threshold sensitivity to free chlorine. A total of 48 panelists were surveyed in the West market. The results are summarized in Table 4.26. The average measured consumer sensitivity to free chlorine was 0.59 mg/L Cl_2 (Figure 4.14). There was less dominance in the higher range of sensitivity thresholds than in the other free chlorine markets. Only 36% of the panelists had a measured sensitivity level greater than 1.0 mg/L Cl_2 (the residual concentration at the first customer's tap), the majority of which were significantly higher (> 1.3 mg/L Cl_2) than the effluent residual.

Table 4.26
Measured threshold sensitivities of TW, BW, FW, and the group as a whole (all) to free chlorine - West market panelists

User group	# of panelists	Log avg. Threshold	Avg. threshold mg/L cl_2	Log S.D.	Max	Min	Median
TW	28	-0.26	0.55	0.49	2.30	0.11	0.54
BW	5	-0.27	0.53	0.34	1.47	0.18	0.61
FW	15	-0.17	0.68	0.46	2.30	0.13	1.00
All	48	-0.23	0.59	0.46	2.30	0.11	0.67

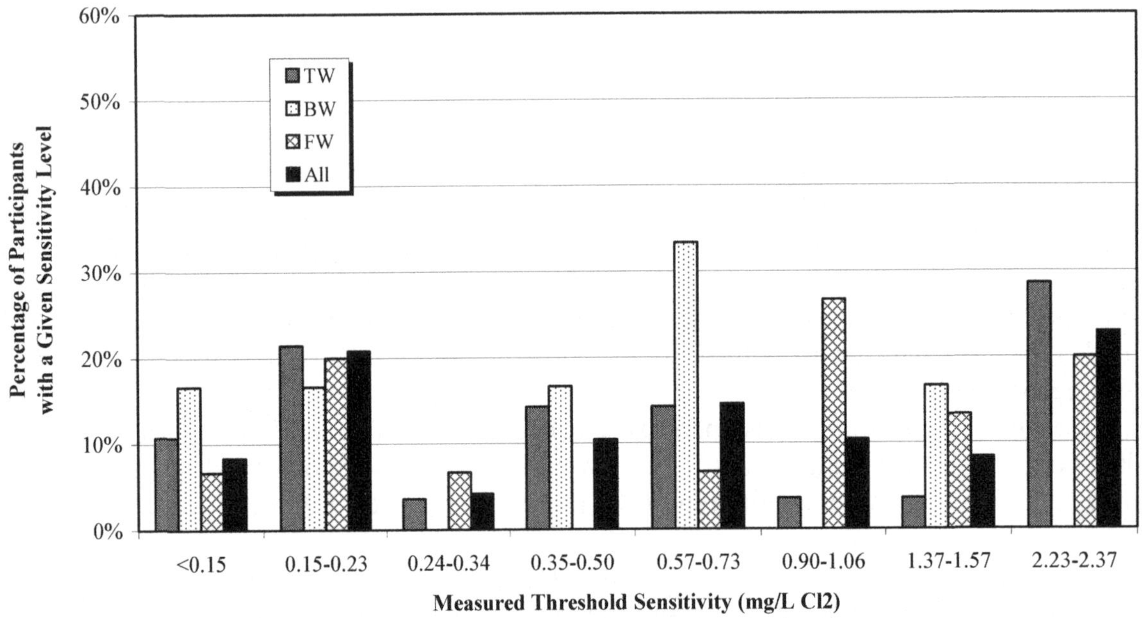

Figure 4.14 Distribution of measured threshold sensitivities to free chlorine - West market panelists

Combined Chlorine Markets

Southeast 2 Market

Consumer perception of water quality. A summary of consumers' responses to questions about their household tap water quality, taste, appearance, smell, safety, and healthiness is provided in Table 4.27. Consumers were also asked to describe the taste and smell of their household tap water; responses are illustrated in Figures 4.15 and 4.16.

Table 4.27
Consumer perception of water quality - Southeast 2 market survey

	Very satisfied (%)	Satisfied (%)	Neither dissatisfied nor satisfied (%)	Dissatisfied (%)	Very dissatisfied (%)
Overall TW quality	34.7	42.3	1.3	13.3	8.4
Overall TW taste	36.1	32.1	0.5	16.2	15.1
Overall TW appearance	57.8	28.3	1.6	4.8	7.5
Overall TW smell	50.4	33.5	1.3	7.3	7.5
Description of TW safety	Safe	Somewhat safe	Neither unsafe nor safe	Somewhat unsafe	Unsafe
	62.1	21.5	5.6	8.1	2.7
Description of TW healthiness	Healthy	Somewhat healthy	Neither unhealthy nor healthy	Somewhat unhealthy	Unhealthy
	51.5	21.8	11.5	10.8	4.4

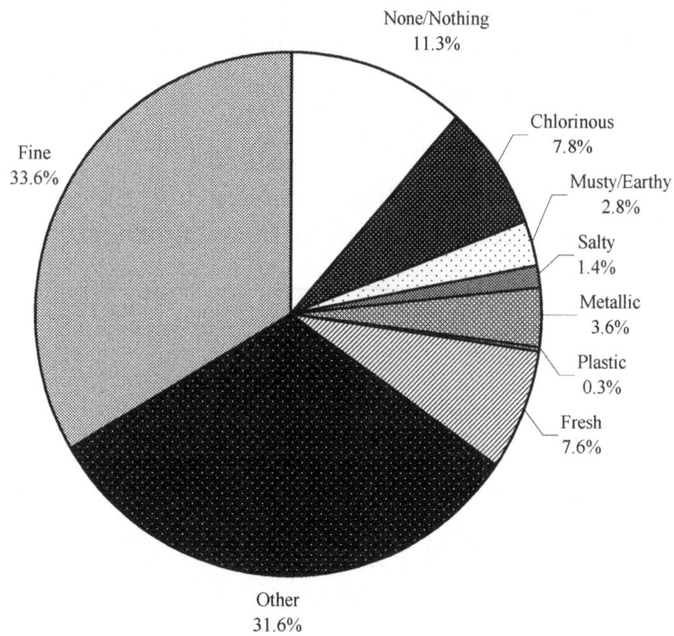

Figure 4.15 Consumer perception of tap water taste - Southeast 2 market survey

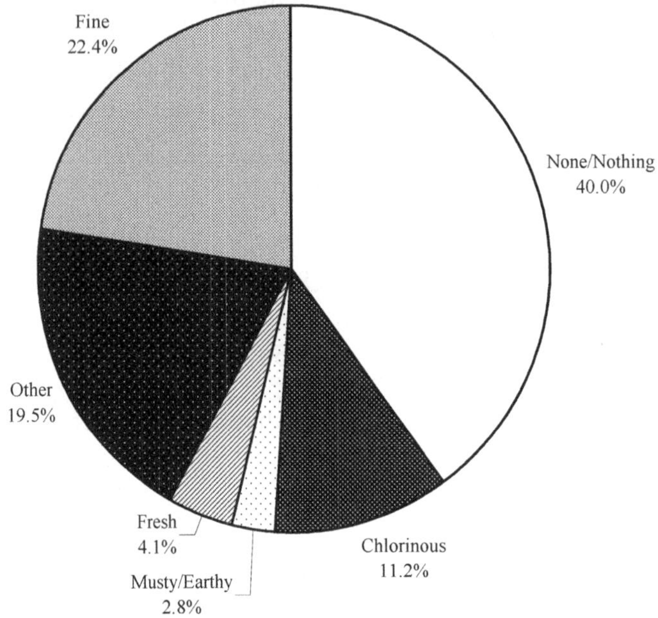

Figure 4.16 Consumer perception of tap water smell - Southeast 2 market survey

Perception of tap water safety. The great majority of respondents were confident their tap water met drinking water regulatory requirements (Table 4.28), though one quarter were not. The majority perceived their tap water to be safe to drink and scientists' and physicians' understanding of tap water-related health risks to be good, though consumers were significantly split on their perception of the safety of chlorinous flavors (Table 4.29).

Table 4.28
Consumer responses to the question "How confident are you that the tap water you receive in your household meets or exceeds state and federal requirements?" - Southeast 2 market survey

Response	% of respondents
Very confident	24.8
Confident	46.1
Neither not confident nor confident	12.1
Not confident	9.9
Not very confident	7.1

Table 4.29
Consumer responses to the statements listed below - Southeast 2 market survey

	"In general, I believe it is safe to drink tap water." (%)	"I believe it is safe to drink water that has a chlorinous taste or odor." (%)	"Scientists and medical experts have a good understanding of any health risks from drinking tap water." (%)
Strongly agree	21.2	4.1	12.9
Agree	51.3	25.8	53.1
Neither disagree nor agree	11.7	12.3	17.9
Disagree	10.1	40.4	12.8
Strongly disagree	5.7	17.4	3.2

Primary drinking water source. Consumers were asked whether they primarily drink tap water or a tap water alternative while at home. Forty-four percent of Southeast 2 market consumers surveyed indicated that they drink primarily tap water in their household, leaving over half of consumers who primarily drink some form of TWA. Among the filtered water drinkers in the Southeast 2 market, 45% said they drink filtered water mainly for overall quality, while 25, 15, and 10% indicated they drink filtered water primarily for health, taste, and safety reasons, respectively. One-third of the bottled water drinkers in the Southeast 2 market indicated that taste was the main reason they chose to drink bottled water, while 18, 16, and 14% said they drink bottled water primarily due to health, safety, and overall quality factors, respectively.

Consumer education. Consumers were asked whether they had received a CCR within the last year. Table 4.30 summarizes their answers. Approximately one-third of the population recalled receiving their water quality report. Of those who received it, most found it useful.

Table 4.30
Consumer awareness of their CCR - Southeast 2 market survey

	Yes (%)	No (%)
Received CCR in the last year	35.6	64.4
Report was useful	78.1	21.9

Threshold sensitivity to monochloramine. A total of 33 panelists were surveyed in the Southeast 2 market. The results are summarized in Table 4.31. The average measured consumer sensitivity to monochloramine was 4.12 mg/L Cl_2 (Figure 4.17). However, the distribution of sensitivities likely had a much greater spread than is indicated by the calculated threshold sensitivities for the panelists (discussed below).

Table 4.31
Measured threshold sensitivities of TW, BW, FW, and the group as a whole (all) to monochloramine - Southeast 2 market panelists

User group	# of panelists	Log avg. threshold	Avg. threshold (mg/L cl_2)	Log S.D.	Max	Min	Median
TW	16	0.61	4.09	0.14	4.95	1.41	4.68
BW	13	0.60	4.00	0.11	4.95	2.09	4.58
FW	4	0.67	4.66	0.00	4.66	4.66	4.66
All	33	0.61	4.12	0.12	4.95	1.41	4.66

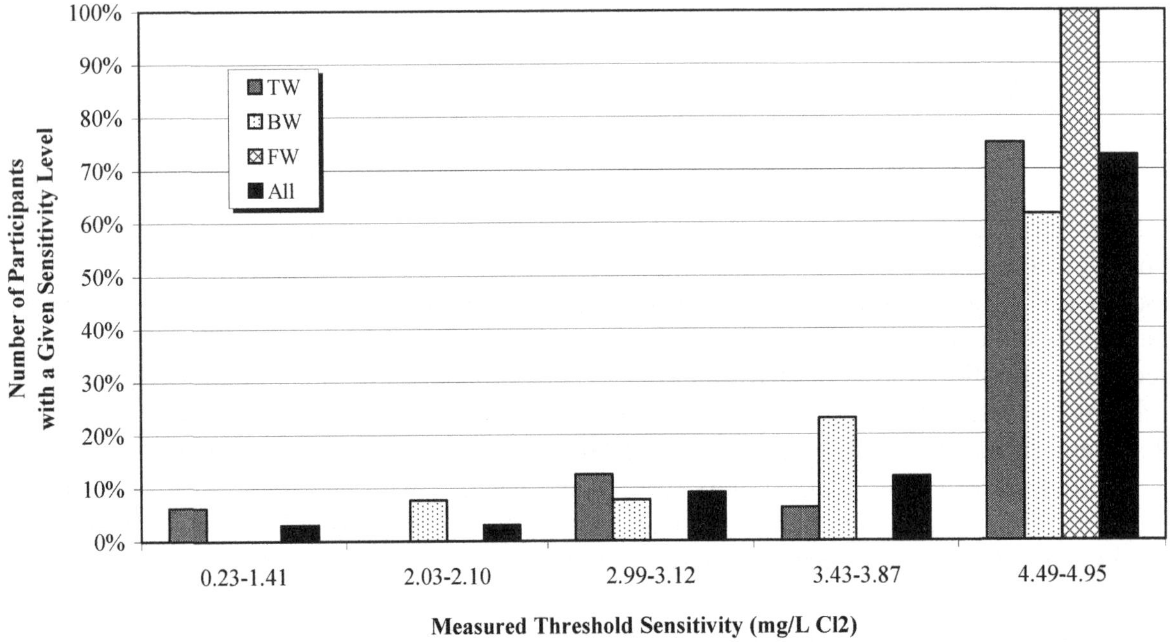

Figure 4.17 Distribution of measured threshold sensitivities to monochloramine - Southeast 2 market panelists

Ideally, "the concentration levels of the test substance in a medium should end at (or above) the concentration at which all panelists give a correct response."[2] The MRDL for monochloramine (4.0 mg/L Cl_2) limited the highest concentration that could be tested. Also, panelists will become saturated after tasting a certain number of samples. When a panelist misses all the samples, as was the case for the majority (73%) of panelists in the Southeast 2 market, their calculated threshold is the geometric average between the highest concentration tested and the next concentration that would have followed in the geometric progression (ASTM Method E679-91). This resulted in a skewing of most of the panelists into one box: 4.49 - 4.95 mg/L Cl_2. The test assumes the panelist would have gotten the next sample in the series. Given the great majority of panelists that missed the highest concentration, it is possible that a significant

[2] ASTM E679-91 (reapproved 1997), Section 6.1

number of them had a threshold sensitivity higher than is predicted by this method (i.e., they would have missed the next sample up too, were it tested).

Seventy-three percent of the panelists had a measured sensitivity level greater than 4.0 mg/L Cl_2, the residual concentration at the first customer's tap. The lowest sensitivity measured was 1.41 mg/L Cl_2 and the highest was 4.95 mg/L Cl_2.

Midwest 2 Market

Consumer perception of water quality. A summary of consumers' responses to questions about their household tap water quality, taste, appearance, smell, safety, and healthiness is provided in Table 4.32. Consumers were also asked to describe the taste and smell of their household tap water; responses are illustrated in Figures 4.18 and 4.19.

Table 4.32
Consumer perception of water quality - Midwest 2 market survey

	Very satisfied (%)	Satisfied (%)	Neither dissatisfied nor satisfied (%)	Dissatisfied (%)	Very dissatisfied (%)
Overall TW quality	52.0	32.3	0.6	8.4	6.7
Overall TW taste	44.3	39.2	0.6	8.6	7.3
Overall TW appearance	55.0	32.8	2.0	4.8	5.4
Overall TW smell	58.6	30.6	2.0	3.7	5.1
Description of TW safety	Safe	Somewhat safe	Neither unsafe nor safe	Somewhat unsafe	Unsafe
	71.3	15.2	6.0	4.3	3.2
Description of TW healthiness	Healthy	Somewhat healthy	Neither unhealthy nor healthy	Somewhat unhealthy	Unhealthy
	54.4	24.5	10.5	7.1	3.5

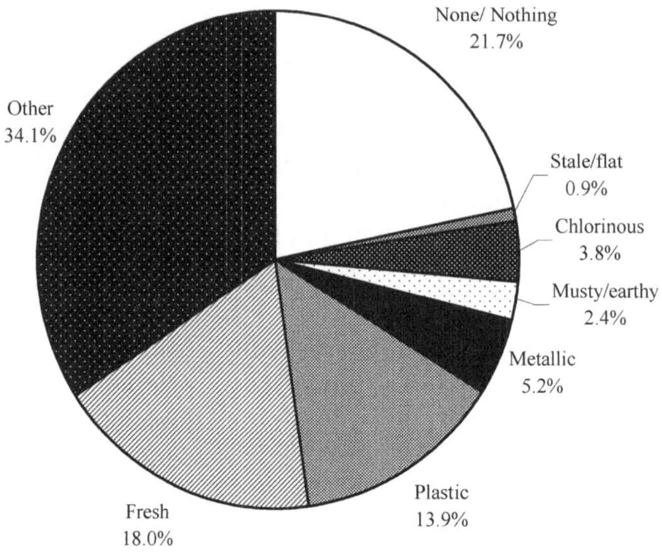

Figure 4.18 Consumer perception of tap water taste - Midwest 2 market survey

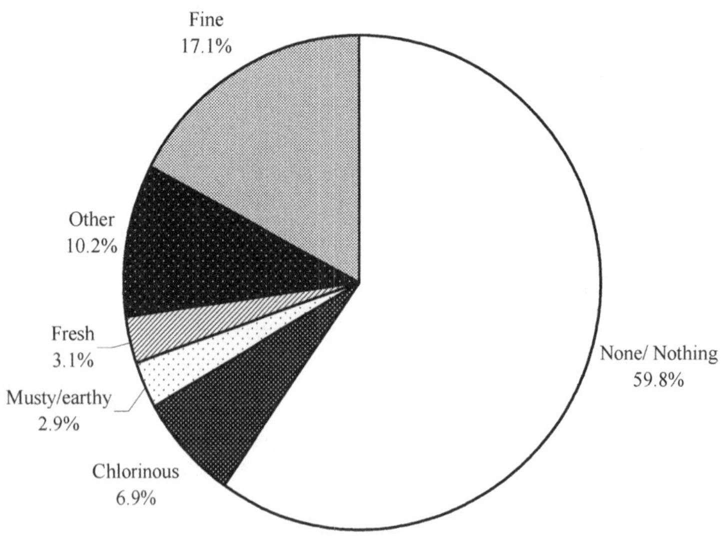

Figure 4.19 Consumer perception of tap water smell - Midwest 2 market survey

Perception of tap water safety. The great majority of respondents were confident their tap water met drinking water regulatory requirements (Table 4.33), though a small fraction, about 10%, were not. The majority perceived their tap water to be safe to drink and scientists' and physicians' understanding of tap water-related health risks to be good, though consumers were significantly split on their perception of the safety of chlorinous flavors (Table 4.34).

Table 4.33
Consumer responses to the question "How confident are you that the tap water you receive in your household meets or exceeds state and federal requirements?" - Midwest 2 market survey

Response	% of respondents
Very confident	25.5
Confident	51.5
Neither not confident nor confident	10.4
Not confident	3.4
Not very confident	9.1

Table 4.34
Consumer responses to the statements listed below - Midwest 2 market survey

	"In general, I believe it is safe to drink tap water." (%)	"I believe it is safe to drink water that has a chlorinous taste or odor." (%)	"Scientists and medical experts have a good understanding of any health risks from drinking tap water." (%)
Strongly agree	17.0	3.0	12.4
Agree	66.2	32.4	64.0
Neither disagree nor agree	6.1	20.3	13.2
Disagree	8.5	29.1	8.3
Strongly disagree	2.3	15.2	2.1

Primary drinking water source. Consumers were asked whether they primarily drink tap water or a tap water alternative (filtered water or bottled water) while at home. Sixty-three percent of Midwest 2 market consumers surveyed indicated that they drink primarily tap water in their household, leaving 37% of consumers who primarily drink some form of TWA. Primary reasons cited for drinking filtered water in the Midwest 2 market were evenly split among taste, health, safety, and overall quality. Twenty-three percent of the bottled water drinkers in the Midwest 2 market indicated that taste was the main reason they chose to drink bottled water, while 21, 19, and 13% said they drink bottled water primarily due to overall quality, health, and safety factors, respectively.

Consumer education. Consumers were asked whether they had received a CCR within the last year. Table 4.35 summarizes their answers. Approximately one-third of the population recalled receiving their water quality report. Of those who received it, most found it useful.

Table 4.35
Consumer awareness of their CCR - Midwest 2 market survey

	Yes (%)	No (%)
Received CCR in the last year	29.8	70.2
Report was useful	84.2	15.8

Threshold sensitivity to monochloramine. A total of 29 panelists were surveyed in the Midwest 2 market. The results are summarized in Table 4.36. The average measured consumer sensitivity to monochloramine was 3.24 mg/L Cl_2 (Figure 4.20). The average threshold here was significantly lower than the other two monochloramine markets (Southwest 1 and West Coast) due to the lower range of concentrations testing in this market. A detailed discussion of the methodology is presented in Chapter 3.

Table 4.36
Measured threshold sensitivities of TW / the group as a whole (all) to monochloramine - Midwest 2 market panelists

User group	# of panelists	Log avg. threshold	Avg. threshold (mg/L cl_2)	Log S.D.	Max	Min	Median
TW/ All	29	0.51	3.24	0.19	3.87	0.59	3.83

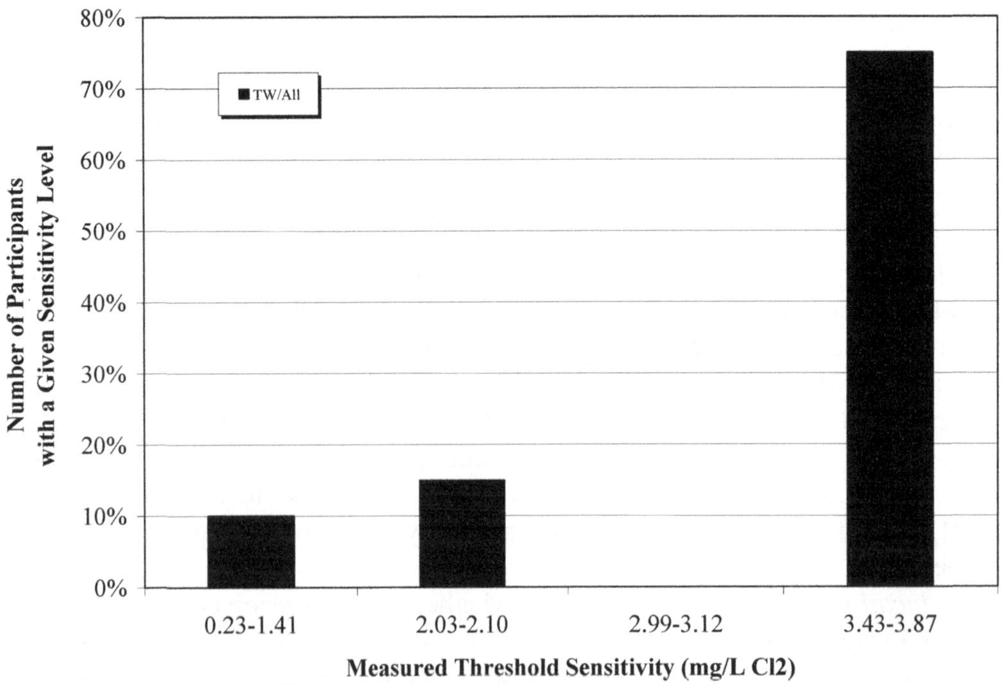

Figure 4.20 Distribution of measured threshold sensitivities to monochloramine - Midwest 2 market panelists

A with the Southwest 2 market, the distribution of sensitivities likely has a much greater spread than is indicated by the calculated threshold sensitivities for the panelists as discussed in detail in the previous section (Southeast 2 market) of this chapter.

Unlike the other six markets tested, due to poor recruiting success in the Midwest 2 market the panelists used here were utility personnel (who did not work around chlorine). As the taste test is a blind test and none of the panelists were trained Flavor Profile Analysts, they fit the general criteria for average consumers. Where they did not is in the subjective portion of the test (the opinion survey), so no comparisons can be made between threshold sensitivity and perception of water quality for this market.

Seventy-five percent of the panelists had a measured sensitivity level greater than 2.3 mg/L Cl_2, the residual concentration at the first customer's tap.

West Coast Market

Consumer perception of water quality. A summary of consumers' responses to questions about their household tap water quality, taste, appearance, smell, safety, and healthiness is provided in Table 4.37. Consumers were also asked to describe the taste and smell of their household tap water; responses are illustrated in Figures 4.21 and 4.22.

Table 4.37
Consumer perception of water quality - West Coast market survey

	Very satisfied (%)	Satisfied (%)	Neither dissatisfied nor satisfied (%)	Dissatisfied (%)	Very dissatisfied (%)
Overall TW quality	49.2	29.4	1.3	12.3	7.8
Overall TW taste	45.4	29.1	1.7	13.1	10.7
Overall TW appearance	64.0	24.2	1.4	6.8	3.6
Overall TW smell	60.2	26.6	2.2	7.1	3.9
Description of TW safety	Safe	Somewhat safe	Neither unsafe nor safe	Somewhat unsafe	Unsafe
	64.3	19.4	3.7	8.3	4.3
Description of TW healthiness	Healthy	Somewhat healthy	Neither unhealthy nor healthy	Somewhat unhealthy	Unhealthy
	56.0	23.0	7.8	8.0	5.2

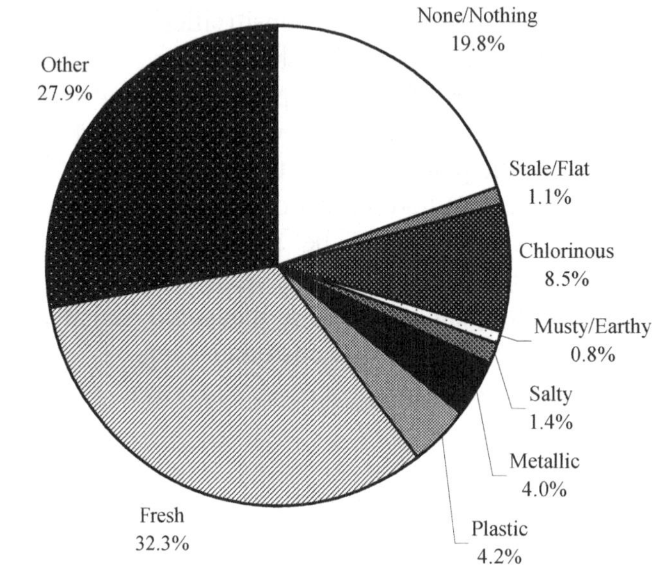

Figure 4.21 Consumer perception of tap water taste - West Coast market survey

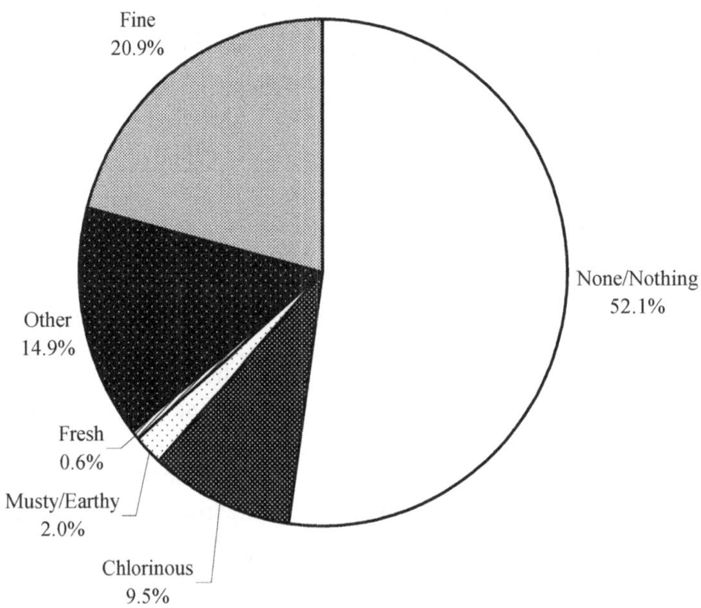

Figure 4.22 Consumer perception of tap water smell - West Coast market survey

Perception of tap water safety. The great majority of respondents were confident their tap water met drinking water regulatory requirements (Table 4.38), though one quarter were not. The majority perceived their tap water to be safe to drink and scientists' and physicians' understanding of tap water-related health risks to be good, though consumers were significantly split on their perception of the safety of chlorinous flavors (Table 4.39).

Table 4.38
Consumer responses to the question "How confident are you that the tap water you receive in your household meets or exceeds state and federal requirements?" - West Coast market survey

Response	% of respondents
Very confident	23.5
Confident	50.6
Neither not confident nor confident	11.7
Not confident	5.5
Not very confident	8.7

Table 4.39
Consumer responses to the statements listed below - West Coast market survey

	"In general, I believe it is safe to drink tap water." (%)	"I believe it is safe to drink water that has a chlorinous taste or odor." (%)	"Scientists and medical experts have a good understanding of any health risks from drinking tap water." (%)
Strongly agree	22.7	6.0	14.6
Agree	49.9	30.6	53.7
Neither disagree nor agree	12.6	13.9	18.0
Disagree	9.3	33.2	10.4
Strongly disagree	5.5	16.3	3.2

Primary drinking water source. Consumers were asked whether they primarily drink tap water or a tap water alternative (filtered water or bottled water) while at home. Forty-six percent of West Coast market consumers surveyed indicated that they drink primarily tap water in their household, leaving over half of consumers who primarily drink some form of TWA. Among the filtered water drinkers in the West Coast market, 42% said they drink filtered water mainly for safety reasons, while 22 and 19% indicated they drink filtered water primarily for taste and health reasons, respectively. Twenty-seven percent of the bottled water drinkers in the West Coast market indicated that taste was the main reason they chose to drink bottled water, while 23, 20, and 14% said they drink bottled water primarily due to safety, health, and overall quality factors, respectively.

Consumer education. Consumers were asked whether they had received a CCR within the last year. Table 4.40 summarizes their answers. Approximately 40% of the population recalled receiving their water quality report. Of those who received it, most found it useful.

Table 4.40
Consumer awareness of their CCR - West Coast market survey

	Yes (%)	No (%)
Received CCR in the last year	40.2	59.8
Report was useful	87.2	12.8

Threshold sensitivity to monochloramine. A total of 30 panelists were surveyed in the West Coast market. The results are summarized in Table 4.41. The average measured consumer sensitivity to monochloramine was 3.64 mg/L Cl_2 (Figure 4.23). However, the distribution of sensitivities likely has a much greater spread than is indicated by the calculated threshold sensitivities for the panelists, as discussed previously in section Southeast 2 of this chapter.

Table 4.41
Measured threshold sensitivities of TW, BW, FW, and the group as a whole (all) to monochloramine - West Coast market panelists

User group	# of panelists	Log avg. threshold	Avg. threshold (mg/L cl_2)	Log S.D.	Max	Min	Median
TW	21	0.54	3.51	0.19	4.60	1.26	4.49
BW	5	0.62	4.18	0.08	4.60	3.03	4.49
FW	4	0.57	3.70	0.17	4.55	2.04	4.48
All	30	0.56	3.64	0.17	4.60	1.26	4.49

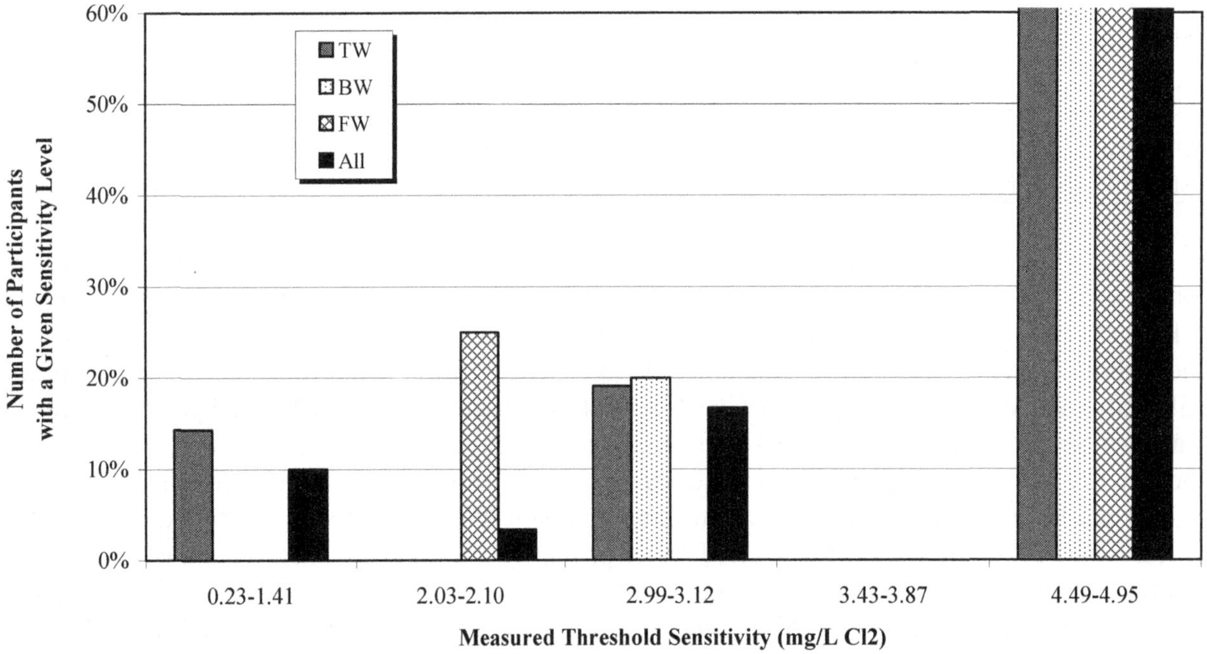

Figure 4.23 Distribution of measured threshold sensitivities to monochloramine - West Coast market panelists

Eighty-seven percent of the panelists had a measured sensitivity level greater than 2.2 mg/L Cl_2 , the residual concentration at the first customer's tap.

DISCUSSION

In this section, market comparisons of TWA drinkers based on demographics, satisfaction levels, and utility characteristics are examined (i.e., are there significant differences in level of satisfaction in terms of both aesthetics and safety/health between demographic groups in the surveyed target markets?). All differences between consumer groups were tested using one-way ANOVA tests or Chi-squared analyses at a 95% level of confidence. The discussion is grouped into three sections:

- Correlations between satisfaction and demographic group.
- Differences in satisfaction with tap water quality between free chlorine and combined chlorine markets.
- Correlation of differences in tap water taste and odor to consumer satisfaction.

Correlations Between Satisfaction and Demographic Group

In order to derive trends within the surveyed consumers, the impacts of age, gender, number of children in the household, and use of tobacco products on the overall rating of satisfaction concerning taste, smell and overall water quality were assessed. In general, gender had no bearing on the choice of the most important water quality parameter, but there were some statistically significant differences in consumer perceptions between some groups within markets (age, tobacco habits, and presence of children).

General differences in satisfaction with tap water quality. In the West Coast market, consumers 65 and older were more likely to indicate appearance as the most important water quality parameter. Additionally, households with children were more likely to choose taste as the most important parameter. In addition to taste, households with children in the Midwest 1 market also mentioned safety. In the East Coast market, consumers 30 to 44 years old were more impacted by healthiness, while consumers 45 to 64 years old chose taste as the critical water quality parameter. Tobacco users were less concerned with safety in the Southwest 1 market.

Satisfaction with tap water taste. Age was the only demographic that showed any statistical difference in satisfaction with tap water taste among consumers. In the West Coast and Midwest 2 markets, average consumers[3] under the age of 30 were less satisfied than consumers 65 and older. Average consumers under the age of 44 in the Southeast 1 market were less satisfied with tap water taste than consumers 65 and older. These data suggest that younger consumers: 1) were more sensitive to tap water taste; 2) had different taste standards; or 3) perceived tastes differently due to valid or invalid personal knowledge (e.g., actual tap water quality, media events). Gender, children, and tobacco use had no bearing on satisfaction with tap water taste.

Satisfaction with tap water smell. Demographic differences had almost no effect on level of satisfaction with tap water smell. The only statistical difference in satisfaction with tap water smell among demographic groups was observed in the Southeast 1 market, where average households with children were more satisfied than average households without children. Age, gender, and tobacco use had no bearing on satisfaction with tap water smell.

[3] The market survey data for each demographic group was compared to the other demographic subgroups (e.g., different age groups) to look for differences in level of satisfaction. The "average" consumer represents the opinion of the average or typical person in that subgroup as measured by the survey.

Differences in Satisfaction with Tap Water Quality between Free Chlorine and Combined Chlorine Markets

Average consumer satisfaction with respect to overall water quality, taste, smell, safety and healthiness is shown in Table 4.42 for the chlorine and combined chlorine target markets and the national market. Most consumers were satisfied with their tap water, although taste was the cause of higher consumer dissatisfaction and was often associated with a perception of tap water unhealthiness and lack of safety. There were no statistically significant differences in satisfaction between the chlorine markets as a group and the monochloramine markets as a group, though there are significant differences in satisfaction between individual markets. This is in line with survey results reported elsewhere (Mackey et al. 2003). In general, greater consumer dissatisfaction was observed in the national survey (as compared to the target markets) for all parameters measured.

Table 4.42
Summary of telephone survey results for the national survey and market surveys - Satisfaction with water quality

		Free Cl_2				NH_2Cl			Nat'l
		EC (%)	SE1 (%)	MW1 (%)	West (%)	SE2 (%)	WC (%)	MW2 (%)	(%)
Overall Satisfaction:	Sat./Very sat.	81	66	82	74	77	79	84	72
	Dissat./Very dis.	17	33	17	24	22	20	15	26
Taste:	Sat./Very sat.	71	65	78	67	68	75	84	66
	Dissat./Very dis.	28	34	23	32	31	24	16	33
Smell:	Sat./Very sat.	85	78	89	80	84	87	89	80
	Dissat./Very dis.	14	21	9	16	15	11	9	17
Safety:	Sat./Very sat.	86	80	85	84	84	84	87	82
	Dissat./Very dis.	10	16	10	8	11	13	8	15
Health:	Sat./Very sat.	74	63	73	77	73	79	79	68
	Dissat./Very dis.	18	21	10	16	15	13	11	21

Table 4.43 summarizes consumer perception of tap water safety with respect to perception of general safety, safety of chlorinous flavors and understanding of tap water-related health risks. As with satisfaction level, there were no statistically significant differences in satisfaction between the free chlorine markets as a group and the monochloramine markets as a group, though there are significant differences in satisfaction between individual markets. The

majority of consumers believed tap water is safe to drink, though the fraction varied from 68 to 83% between survey groups. Though few consumers (16% or less in all groups) felt that scientists and medical experts do not have a good understanding of tap water-related health risks, there was a significant variation in agreement (62 - 81%).

Table 4.43
Summary of Telephone Survey results for the national survey and market surveys - Perception of tap water safety

		Free Cl_2				NH$_2$Cl			
		EC (%)	SE1 (%)	MW1 (%)	West (%)	SE2 (%)	MW2 (%)	WC (%)	Nat'l (%)
"In general, I believe it is safe to drink tap water."	Agree/Strongly Agree	72	67.9	73.5	82.8	72.5	83.2	72.6	72.5
	Disagr./Strongly Disagr.	14.7	18.6	15.8	13.1	15.8	10.8	14.8	15.8
"I believe it is safe to drink water that has a chlorinous taste or odor."	Agree/Strongly Agree	37	34.4	35.9	53.4	29.9	35.4	36.6	29.9
	Disagr./Strongly Disagr.	53.9	54.5	47.8	40.5	57.8	44.3	49.5	57.8
"Scientists and medical experts have a good understanding of any health risks from drinking tap water."	Agree/Strongly Agree	61.9	69.3	75.2	80.8	66	76.4	68.3	66.0
	Disagr./Strongly Disagr.	16	11.5	11.6	8.9	16	10.4	13.6	16.0

Perception of chlorinous taste or odor was generally negative, though the fraction who viewed it unsafe ranged from 40.5 to 57.8% between survey groups. In only one market (West) did the majority of consumers perceive a chlorinous taste or odor to be safe.

This survey gives a good general sense of consumer satisfaction with tap water quality and of chlorinous flavors - the majority of consumers feel tap water is safe to drink, though there is much distrust of chlorinous tastes and odors. However the significant variation in some parameters from market to market (e.g., the fraction of consumers dissatisfied with taste ranges from 15 to 33%) highlights the difficulty a water utilities face when trying to assess their consumers' perception of water quality and what might be influencing their opinion. Local culture, water quality, water temperature, etc., all influence consumer satisfaction with tap water quality and their decisions about whether to purchase TWAs.

Correlation of Differences in Tap Water Taste and Odor to Consumer Satisfaction

In order to assess whether consumers' perceptions of their tap water taste and odor likely affect their perception of tap water safety and healthiness, differences in average safety and healthiness ratings were calculated among consumer groups that categorized their tap water taste or odor in a certain way (i.e., none/fresh/good, chlorinous, bad, other). The "bad" category combined the following responses: stale/flat, musty/earthy, metallic, salty, and plastic. The "other" category incorporated only open-ended responses (i.e., consumers wrote in a description instead of choosing one of the pre-determined categories, which typically included descriptions such as "gross," "nasty", "dirty", "sulfurous", and "obnoxious").

Figure 4.24 illustrates the trend between tap water taste/odor description and average perception of safety for the National Survey. A rating of "1" corresponded to "unsafe," while a rating of "5" corresponded to "safe." As expected, consumers who thought that their tap water had no taste/odor or a fresh taste/odor gave their tap water a higher safety rating than those who felt their tap water had an "other", bad, or chlorinous taste. With an average safety rating near 4, those who described their tap water taste/odor as chlorinous still felt that the water was somewhat safe to drink. Consumers who described their tap water as tasting or smelling stale/flat, musty/earthy, metallic, salty, or plastic were the most concerned about the safety of their tap water. However, in general, the lowest rating obtained (for odor) was still positive (above 3, which corresponded to neutral - "neither safe nor unsafe").

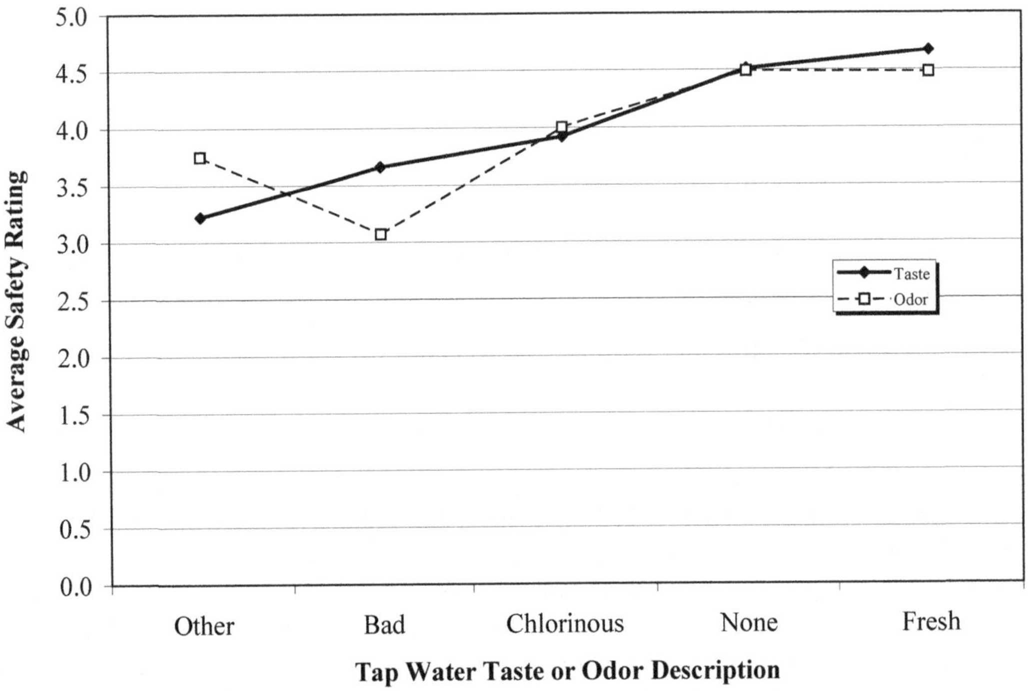

Figure 4.24 Impact of taste and odor description on tap water safety rating, for the National survey

The national survey showed that the general trend observed between tap water taste/odor description and healthiness rating was similar to that observed for the taste/odor description and safety rating (less detection of off flavors translated to higher healthiness ratings), although, average healthiness ratings associated with a taste/odor were lower than average safety ratings (Figure 4.25).

Figures 4.26 and 4.27 illustrate the impact of perceived tap water flavors on the consumer's perception of safety and healthiness for the seven drinking water markets surveyed. In general, the observed individual market trends reflected the national market trends -

perception of fresh taste resulted in a much higher perception of water safety and healthiness. Chlorinous-tasting tap water was generally perceived as safer than tap water with other off flavors. However, in two markets (Midwest 2 and Southeast 1), average consumers who described their tap water as having a chlorinous taste rated the safety and healthiness of their tap water lower than average consumers who described their tap water as having a bad taste. Consumers in these two markets appeared to be particularly concerned about the health implications associated with chlorinous tastes in their drinking water.

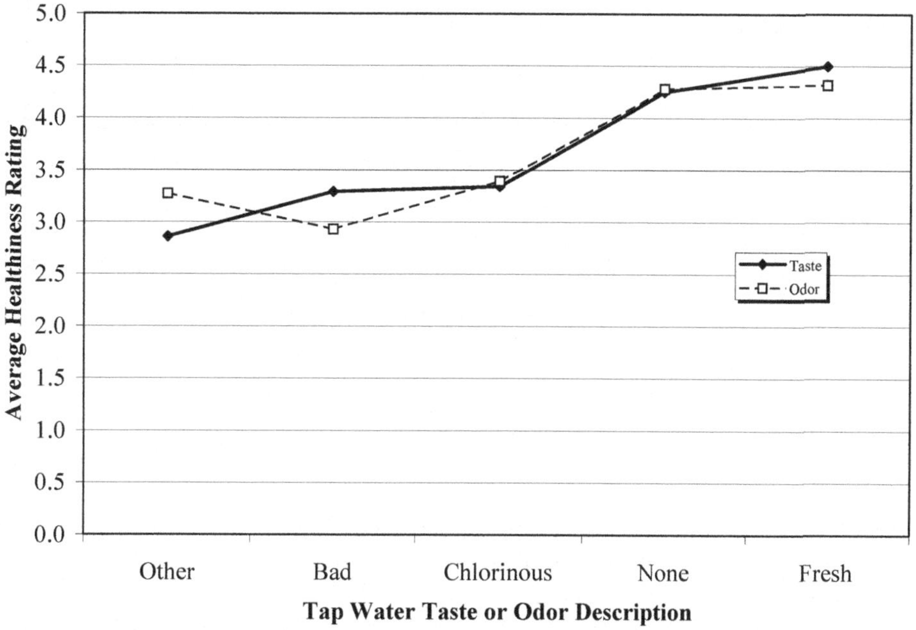

Figure 4.25 Impact of taste and odor description on tap water healthiness rating - National survey

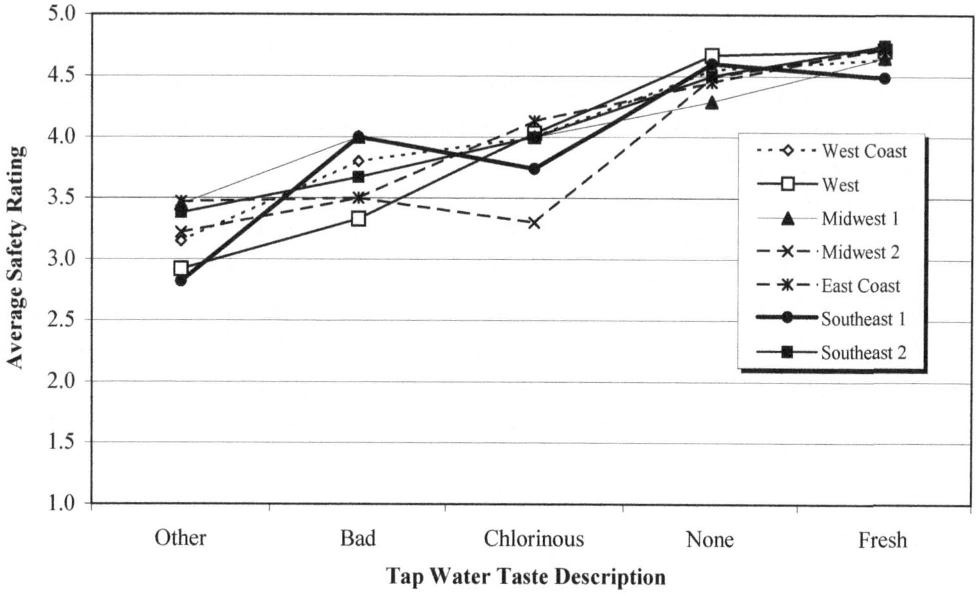

Figure 4.26 Impact of taste description on tap water safety rating, for each individual market

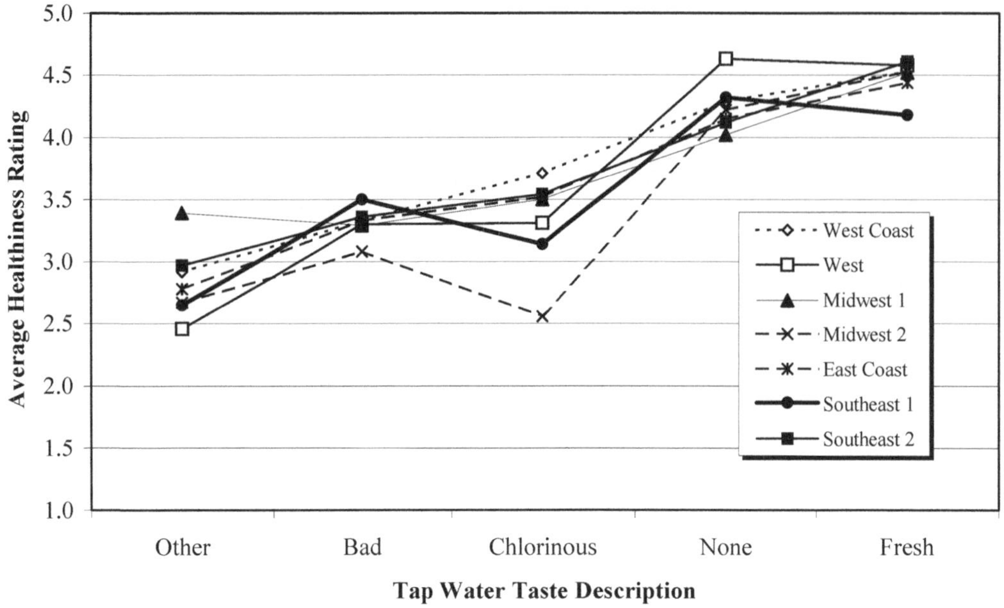

Figure 4.27 Impact of taste description on tap water healthiness rating for each individual market

Trends describing tap water odor description vs. safety and healthiness ratings for individual markets were somewhat similar to those observed for tap water taste description vs. safety and healthiness ratings (Figures 4.28 and 4.29). The only appreciable difference was among consumers who thought their water had a "bad" odor. In the Southeast 2, Midwest 1, West Coast, and East Coast markets, a "bad" odor description was associated with the lowest safety and healthiness ratings. In the West and Midwest 2 markets, "other" odor descriptions were associated with lower safety and healthiness ratings than "bad" odor descriptions. In the West market, consumers were not particularly concerned with "bad" odors at all - the average safety and healthiness ratings given by West market consumers describing their tap water odor as "bad" were 4.5 and 4.0, respectively. No consumers in the Southeast 1 market described the taste or odor of their tap water as "bad."

According to the results collected from the national and regional market surveys, when TWA consumers were asked why they switched from TW to a TWA, the dominant motivator cited was taste. Since "chlorinous taste" was one of the most commonly cited taste descriptors, it was hypothesized that increased chlorinous taste would motivate consumers to switch to TWAs. To examine this hypothesis, the percentage of consumers that described their tap water taste as "chlorinous" was related to the percentage of TWA users, Figure 4.30. There was no obvious trend in selection of TWAs as a function of perception of chlorinous flavors in tap water.

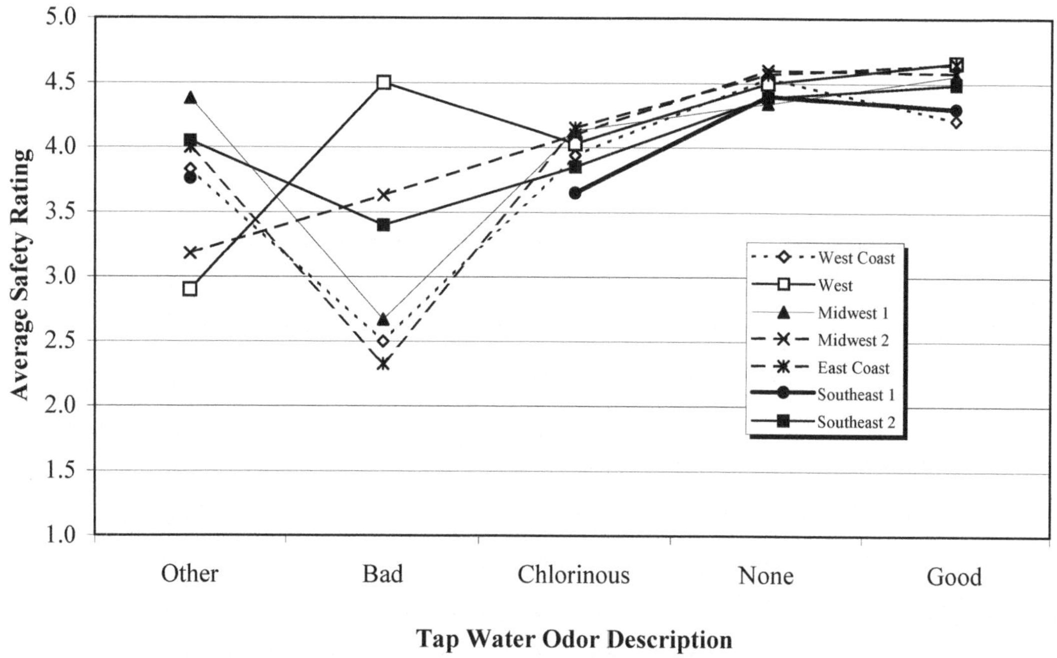

Figure 4.28 Impact of odor description on tap water safety rating, for each individual market

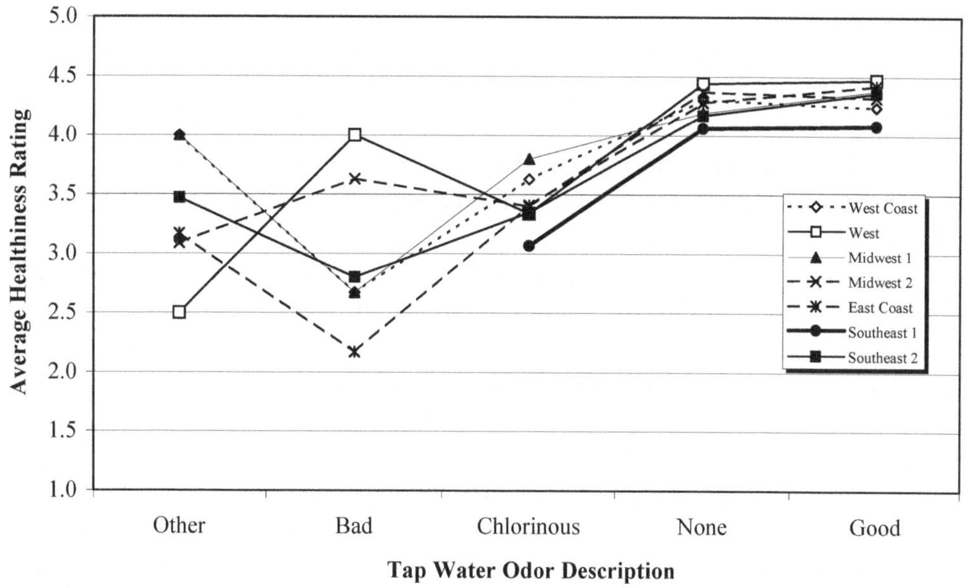

Figure 4.29 Impact of odor description on tap water healthiness rating, for each individual market

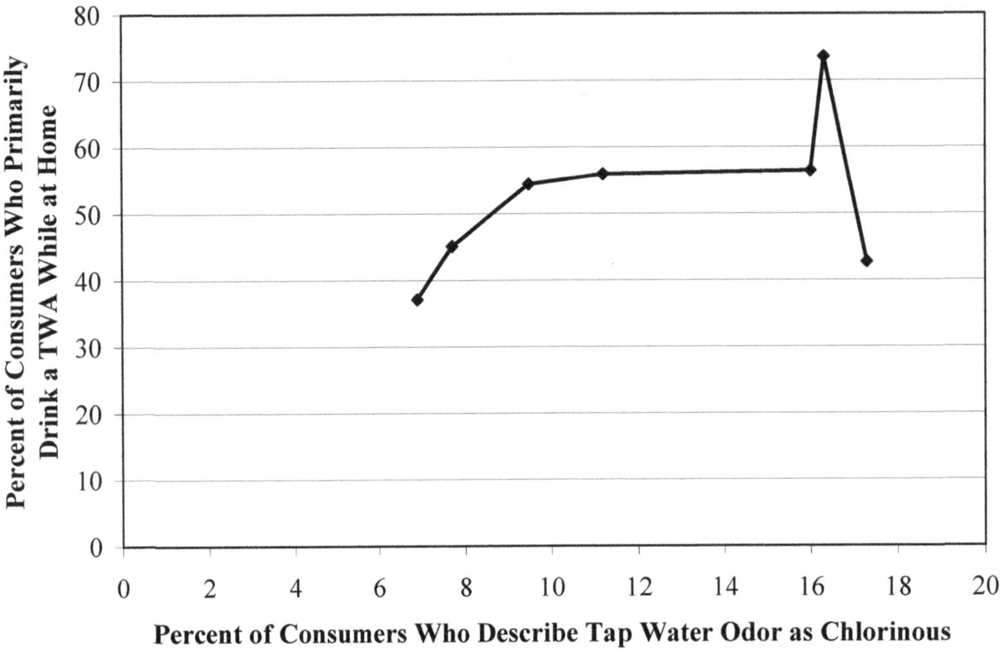

Figure 4.30 Perception of chlorinous odor in tap water versus fraction of consumers using TWAs in the seven surveyed target markets

Between February and April 1998, the West Coast market utility changed their distribution system disinfectant from free chlorine to combined chlorine. Their drinking water quality report from spring 1999 stated that this conversion to combined chlorine greatly reduced the chlorinous taste of its tap water. This is objectively borne out in this study by the great difference in the average consumer sensitivity threshold measured for free chlorine (~ 1 mg/L as Cl_2) versus combined chlorine (> 4 mg/L as Cl_2). To further explore this possible link, differences in consumer satisfaction between markets that use free chlorine and those that use combined chlorine for distribution system disinfection were investigated.

It was found that consumers within free chlorine markets were more likely to describe their tap water taste and odor as chlorinous than consumers within combined chlorine markets, however, no statistical difference was observed between free chlorine market consumers and combined chlorine market consumers in terms of satisfaction with household tap water aesthetics (i.e., taste, smell, and appearance) or in terms of confidence that their tap water meets or exceeds all drinking water standards. Consistent with these data, it was found that 48.9 and 48.1% of combined and free chlorine market consumers, respectively, primarily drink tap water while at home. That is, there is no difference in TWA use between consumers in markets that use free chlorine and markets that use combined chlorine. Therefore, while the use of free chlorine as a distribution system disinfectant may cause increased consumer detection of chlorinous taste and odor, it does not affect average consumer satisfaction with tap water quality nor does it account for trends in TWA usage. While some fraction of consumers will choose TWAs due to chlorinous flavor in the water, there are many factors that contribute to the growing consumption of TWAs over tap water by American consumers. These factors include other tastes and odors (musty/earthy, metallic, medicinal), and perceptions of tap water health and safety. The effect of marketing by purveyors of TWAs could not be quantified in this study.

CHAPTER 5
CONSUMER PERCEPTIONS OF AND SENSITIVITY TO CHLORINOUS FLAVORS IN THEIR TAP WATER

INTRODUCTION

It was hypothesized that individuals with low threshold sensitivity to chlorinous flavors are more likely to have a negative opinion of tap water quality. If this is true, then lowering the residual concentration in the distribution system might significantly improve consumer satisfaction with tap water quality. The complete results of the taste tests presented in this chapter are detailed in Appendix D.

OBJECTIVE

The objective of this chapter is to determine whether chlorinous flavors detection by consumers is supported by their sensitivities to these compounds or mostly driven by perception. Additionally, the impact of consumer sensitivity to chlorinous flavors as compared to their perceptions of water quality is assessed.

RESULTS AND DISCUSSION

Consumer Sensitivity to Chlorinous Flavors

Free Chlorine

Average threshold sensitivities to free chlorine for each market tested and the combined results for the four free chlorine markets are summarized in Table 5.1. The combined average measured consumer sensitivity to free chlorine was 0.82 mg/L Cl_2, approximately three times the value reported in the literature for a trained panel (0.24 - 0.30 mg/L Cl_2, Krasner and Barret 1984). Only 20% of panelists had a threshold sensitivity at or below the literature value.

Table 5.1

Panelist threshold sensitivity to free chlorine (mg/L Cl_2) in their tap water

Location	User type	#	Avg.	+S.D.	-S.D.	Max	Min
East Coast	Total	41	0.77	2.01	0.30	2.31	0.12
	TW	20	0.62	1.88	0.20	2.31	0.12
	BW	8	0.91	1.64	0.51	2.25	0.40
	FW	13	0.99	2.27	0.43	2.28	0.19
Southeast 1	Total	40	1.10	2.49	0.48	2.37	0.13
	TW	16	1.27	2.80	0.58	2.34	0.19
	BW	12	0.83	2.17	0.32	2.37	0.13
	FW	12	1.19	2.38	0.59	2.37	0.38
Midwest 1	Total	30	1.02	2.94	0.35	2.36	0.09

(continued)

Table 5.1 (continued)

Location	User type	#	Avg.	+S.D.	-S.D.	Max	Min
	TW	14	1.09	3.43	0.34	2.36	0.09
	BW	8	1.21	1.8	0.81	2.28	0.72
	FW	8	0.76	3.04	0.19	2.36	0.09
West	Total	48	0.59	1.69	0.20	2.30	0.11
	TW	28	0.55	1.70	0.18	2.30	0.11
	BW	5	0.53	1.16	0.25	1.47	0.18
	FW	15	0.68	1.95	0.24	2.30	0.13
All Markets	Total	159	0.82	2.22	0.30	2.37	0.09
	TW	78	0.62	1.88	0.20	2.31	0.12
	BW	33	0.87	1.85	0.41	2.37	0.13
	FW	48	0.88	2.34	0.33	2.37	0.09

As would be expected with a randomly selected human population, there was a large variation in sensitivities (Figure 5.1). Forty-six percent of the panelists had a measured sensitivity level greater than the highest plant residuals (at the tap) used in their respective markets. This would indicate that these consumers were unable to detect free chlorine in their household tap water, as the residual concentration had to be at or below the plant effluent concentration.

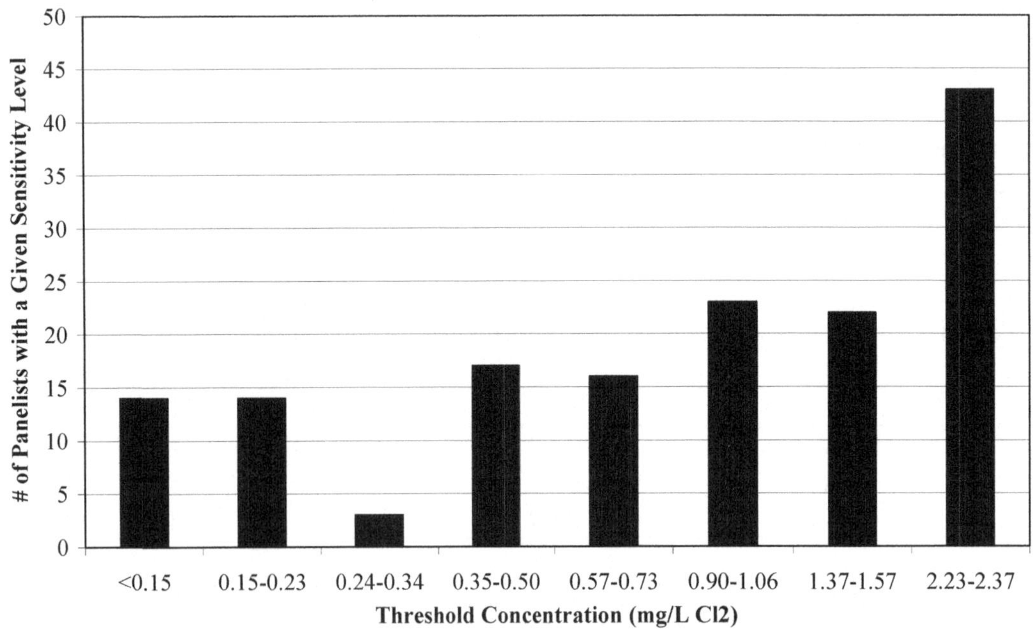

Figure 5.1 Threshold of free chlorine sensitivity distribution, detected by consumers in free chlorine markets

There was a significant difference between the lowest and highest average thresholds (0.59 and 1.1 mg/L Cl_2, respectively) and the distribution of thresholds between markets (Figure 5.2).

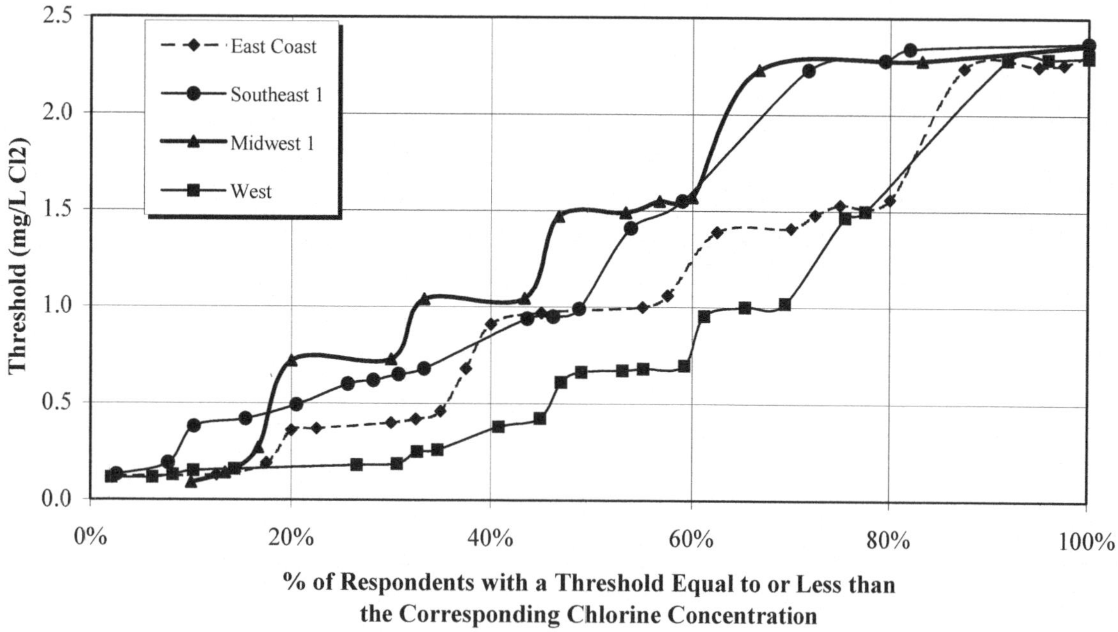

Figure 5.2 Percent of individuals with a measured free chlorine threshold sensitivity greater than or equal to the corresponding value

The lowest sensitivity measured was 0.09 mg/L Cl_2 and the highest was 2.37 mg/L Cl_2. However, the high end of the distribution is very likely greater than is indicated by the calculated threshold sensitivities for the panelists. As discussed in Chapter 4, ideally, "the concentration levels of the test substance in a medium should…end at (or above) the concentration at which all panelists give a correct response."[1] When a panelist misses all the samples, as was the case for a few of the panelists in the free chlorine markets, the calculated threshold is the geometric average between the highest concentration tested, and the next concentration that would have followed in the geometric progression (approximately 2.85 mg/L Cl_2 in this case). While this is the "best estimate" of the panelists' threshold, it is not unlikely that some of the panelists had thresholds substantially higher than this value.

Table 5.2 summarizes the perceptions of all consumers surveyed in each market and compares these results to those of the taste test panelists. Consumer perceptions were compared between those who reported chlorinous taste and/or odors (CLTO) and chlorinous odors only (CLO) and those whose threshold sensitivities were higher or lower than 1.0 mg/L. Additionally, the water quality perception between consumers with sensitivity thresholds

[1] ASTM E679-91 (reapproved 1997), Section 6.1

Table 5.2

Average perceptions of market survey participants and taste test panelists as a function of sensitivity to free chlorine relative to a 1.0 mg/L Cl_2 free chlorine residual. (Market residuals varied between 0.8 and 1.0 mg/L Cl_2)

	All surveyed markets	Taste test panelists	Reported*		Threshold (mg/L Cl_2)*		Threshold > 1.0 mg/L Cl_2‡			Threshold ≤ 1.0 mg/L Cl_2§		
			CLTO†	CLO†	> 1.0	≤ 1.0	CLTO	Other	G/N	CLTO	Other	G/N
% of panelists in category			34%	11%	46%	54%	33%	23%	44%	36%	24%	40%
Overall quality	3.85	3.58	3.19	3.44	3.61	3.55	3.33	3.18	4.06	3.07	3.30	4.12
Taste	3.67	3.36	2.70	3.06	3.41	3.32	2.79	2.76	4.22	2.63	3.10	4.06
Smell	4.18	3.62	3.00	3.17	3.67	3.58	2.96	3.59	4.26	3.03	3.32	4.15
Appearance	4.41	3.85	3.61	3.83	3.90	3.81	3.75	3.65	4.16	3.50	3.68	4.21
Safety	4.31	4.12	4.04	4.20	4.08	4.16	3.95	3.79	4.31	4.13	4.16	4.19
Healthiness	4.00	3.85	3.78	3.92	3.83	3.87	3.78	3.50	4.04	3.78	3.50	4.17
Safe to drink Cl water	2.82	3.25	3.25	3.12	3.09	3.39	3.00	2.87	3.24	3.43	2.82	3.64
Residual	0.8 - 1.0 mg/L Cl_2											

* Panelists sample number: 157

† CLTO: Chlorinous taste and/or odors; CLO: Chlorinous odor only

‡ Sample Number: 73

§ Sample Number: 84

The panelists were divided between those who had reported chlorinous taste and/or odor (CLTO), other (OTH), and Good/None (G/N) when asked in the survey what type of taste was present in their water.

above or lower than 1.0 mg/L was evaluated based on what other tastes were detected (chlorinous, other, and good/none). The sensitivities measured during the taste tests were associated with the consumers' perceptions (reported in the survey) and the average water quality ratings were compared. According to the results shown:

- 46% of panelists had a threshold above the residual.
- 34% of panelists reported a chlorinous flavor in their water; 57% (31/54) of whom had thresholds below the highest free chlorine residual in their market (at the entrance to the distribution system).
- Of those who perceived a chlorinous odor (much lower threshold), 2/3 also reported a chlorinous taste.
- There was no significant difference in tap water perception of aesthetics or safety/health between those consumers with threshold below and above the chlorine residual.
- There was no significant difference in perception of chlorinous flavors between panelists with a low threshold sensitivity and those with a high threshold sensitivity - there was no correlation between perception of chlorinous flavor and measured ability to detect it at levels commonly found in tap water.
- As with the market survey consumers, taste test panelists who perceived no/fresh/good tap water flavors were more satisfied with their water quality at all levels, regardless of their threshold sensitivity.

Monochloramine

Average threshold sensitivities to monochloramine for each market are summarized in Table 5.3. Additionally, the average for all surveyed markets are also included. The average measured consumer sensitivity to monochloramine was 3.67 mg/L Cl_2, more than seven times the value reported in the literature for a trained panel (0.28 - 0.48 mg/L Cl_2, Krasner and Barret 1984 and Piriou and Pirelle 1999). In fact, the literature value threshold was very close to the minimum threshold measured in the combined chlorine markets (0.59 mg/L Cl_2).

Table 5.3

Panelist sensitivity to monochloramine (mg/L) in their market's tap water

Location	User type	#	Avg.	+S.D.	-S.D.	Max	Min
Southeast 2	Total	33	4.12	5.44	3.12	4.95	1.41
	TW	16	4.09	5.67	2.94	4.95	1.41
	BW	13	4.00	5.18	3.09	4.95	2.09
	FW	4	4.66	4.66	4.66	4.66	4.66
Midwest 2	All/TW	29	3.24	5.02	2.09	3.87	0.59
West Coast	Total	30	3.64	5.43	2.44	4.60	1.26
	TW	21	3.51	5.46	2.25	4.6	1.26
	BW	5	4.18	5.01	3.49	4.6	3
	FW	4	3.7	5.49	2.49	4.55	2.04

(continued)

Table 5.3 (continued)

Location	User type	#	Avg.	+S.D.	-S.D.	Max	Min
All Markets	Total	92	3.67	5.38	2.50	4.95	0.59
	TW	66	3.67	5.36	2.51	4.95	0.59
	BW	18	3.89	5.12	2.96	4.95	2.03
	FW	8	3.80	5.23	2.76	4.95	2.03

As would be expected with a randomly selected human population, there was a large variation in sensitivities (Figures 5.3 and 5.4). Eighty-seven percent of the panelists had a measured sensitivity level greater than 2.5 mg/L Cl_2, the highest plant effluent concentration used in the majority of monochloramine markets. This strongly indicates that most consumers were unable to detect combined chlorine in their household tap water.

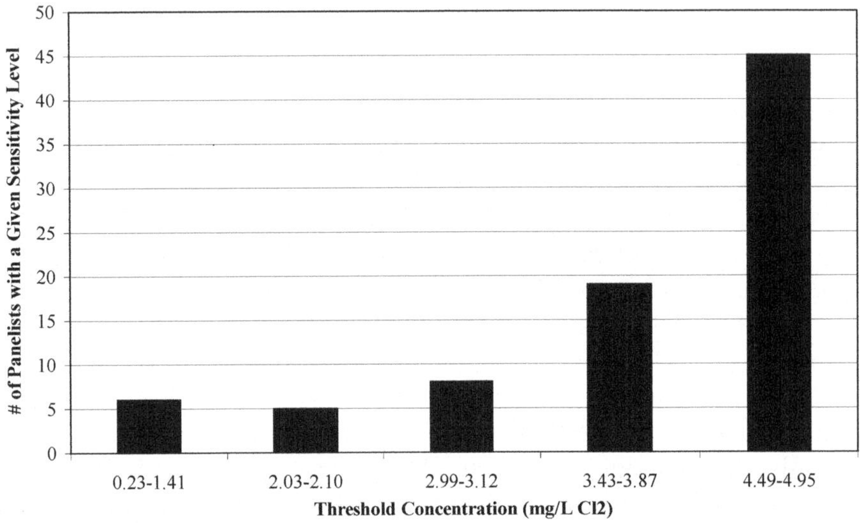

Figure 5.3 Distribution of measured threshold sensitivities to monochloramine for panelists in the combined chlorine markets

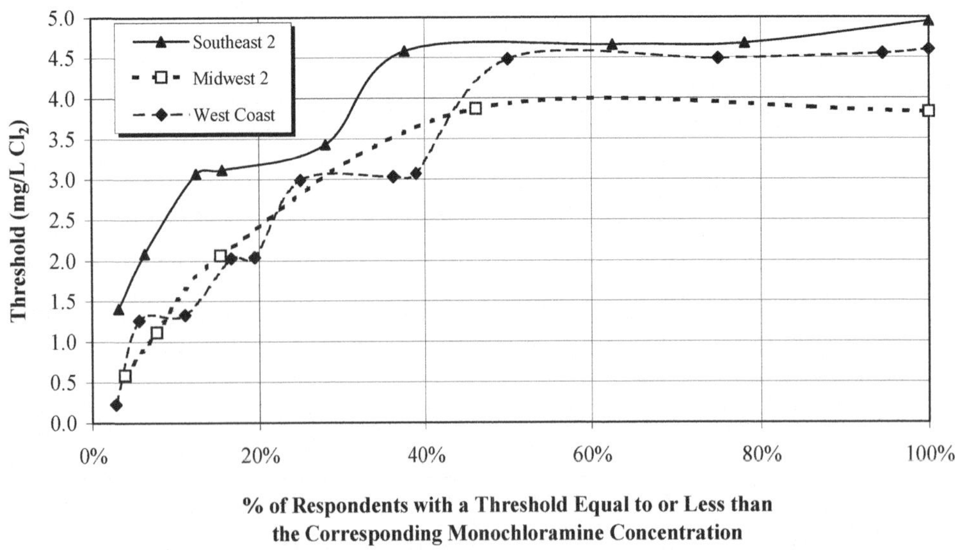

Figure 5.4 Percent of individuals with a measured monochloramine threshold sensitivity greater than or equal to the corresponding value

As discussed in Chapter 4, the data strongly suggests that the average consumer sensitivity was significantly underestimated as: 1) at the first site, Midwest 2, the range of concentrations was based on the literature value of 0.48 mg/L Cl_2, which turned out to grossly underestimate the average consumer sensitivity threshold; 2) the highest concentration that could be tested in the second two markets (Southwest 2 and West Coast) was limited by the MCL for monochloramine (4 mg/L Cl_2); 3) even though the concentrations were increased at the following two sites and the great majority of panelists continued to miss all of the samples. When all of the samples are missed, it is assumed that the panelist would have detected the next concentration that would have followed in the geometric progression. Given that over 75% of the panelists missed all the samples tested, it is likely that the measured threshold sensitivity for some of the panelists was underestimated.

Sensitivity to Monochloramine

Table 5.4 summarizes the perceptions of all consumers surveyed and compares them to those of the taste test panelists. Table interpretation is the same as for Table 5.2. A similar analysis to the one performed for the chlorinous markets was conducted for the monochloramine market:

- The great majority of the panelists (87%) were not sensitive to monochloramine residuals of 2.5 mg/L Cl_2.

  This assessment did not include the panelists pertaining to the MW2 market. Due to consumer recruiting difficulties, utility personnel were used for the average consumer taste tests, and therefore, were ineligible to complete the survey. However, including their results would not have made a great impact in what concerns sensitivity to chloramine residual (85% instead of 87% would have a threshold above 2.5 mg/L Cl_2).
- As in the free chlorine markets, there were no differences in consumer perception of tap water aesthetics and safety/health between panelists with threshold sensitivities above and threshold sensitivities below 2.5 mg/L Cl_2.
- As with the market survey consumers, taste test panelists who did not perceive any "off" or bad flavors (including chlorinous flavors) were more satisfied regardless of their threshold sensitivity.
- There was no significant difference in perception of chlorinous flavor in tap water between the high threshold and low threshold (who, presumably, are much more likely to taste it) panelists - there was no correlation between perception of chlorinous flavor and measured ability to detect it at levels commonly found in tap water;

These results imply that consumers often do not accurately identify chlorinous tastes and/or odors they perceive in their tap water. This is investigated further in Chapter 6. It is speculated that consumers may misidentify some objectionable odors as "chlorinous" because that is one drinking water constituent most people know is present in the water.

Table 5.4

Average perceptions of market survey participants and taste test panelists as a function of sensitivity to monochloramine relative to a 2.5 mg/L Cl_2 monochloramine residual (market residuals varied between 2.2 and 4.5 mg/L as Cl_2)

Average perception	All surveyed markets	Taste test panelists	Reported*		Threshold (mg/L Cl_2)*		Threshold (mg/L Cl_2) > 2.5‡			Threshold (mg/L Cl_2) < 2.5§		
			CLTO†	CLO†	> 2.5	≤ 2.5	CLTO	Other	G/N	CLTO	Other	G/N
% of panelists in category:			22%	9%	87%	13%	22%	33%	45%	22%	11%	67%
Overall quality†	3.99	3.71	3.13	3.50	3.65	4.11	3.00	3.35	4.15	4.00	4.00	4.17
Taste†	3.84	3.52	2.40	2.33	3.45	4.00	2.31	3.25	4.11	3.00	4.00	4.33
Smell†	4.26	3.76	2.93	3.20	3.67	4.33	2.75	3.45	4.23	4.00	4.00	4.50
Appearance†	4.30	3.78	3.00	2.67	3.75	4.00	2.85	3.55	4.33	4.00	2.00	4.33
Safety‡	4.36	3.81	3.42	3.33	3.77	4.17	3.27	3.56	4.08	5.00	4.00	4.00
Healthiness§	4.15	3.48	2.75	2.60	3.49	3.43	2.70	3.19	3.96	3.00	3.00	3.75
Safe to drink Cl water	2.71	2.75	2.67	2.83	2.83	2.22	2.77	2.74	2.85	2.00	2.00	2.33
Residual	2.2 - 4.5 mg/L Cl_2											

* Panelists sample number: 69

† CLTO: Chlorinous taste and/or odor; CLO: Chlorinous odor only

‡ Sample Number: 60

§ Sample Number: 9

The panelists were divided between those who had reported chlorinous taste and/or odor (CLTO), other (OTH), and Good/None (G/N) when asked in the survey what type of taste was present in their water.

Effect of Water Quality Parameters on Threshold Sensitivity

The water quality hardness, alkalinity, pH and chlorine residual at the entrance to the distribution system of the target market source waters is detailed in Table 5.5. Market differences in background tap water parameters were speculated to have an effect on consumer satisfaction or sensitivity. For instance, it was hypothesized that increased treated water TDS, hardness or temperature would tend to make consumers less satisfied with their water quality. However, correlations tended to be weak with the highest R^2 found among the combined chlorine markets regarding satisfaction with healthiness (Table 5.6). The R^2 in this case was low ($R^2 = 0.13$), even for a social science study, which might consider an R^2 of 0.5 to 0.7 as being significant. No correlation coefficient is reported for "FC (free chlorine)" and "CC (combined chlorine) markets" and "chlorine sensitivity" due to the fact that these markets cannot be combined when examining chlorine sensitivity levels. A low sensitivity for a CC market may be considered 2 - 3 mg/L Cl_2 but in a FC market, such level would be considered high.

Table 5.5
Water quality parameters measured during the taste tests

Market	Free chlorine residual (mg/L Cl_2)	Total chlorine residual (mg/L Cl_2)	Hardness (mg/L as $CaCO_3$)	Alkalinity (mg/L as $CaCO_3$)	pH
East Coast	1.0		32	20	7.2
Southeast 1	2.0		132	135	8.1
Midwest 1	0.7		138	103	7.3
West	0.8		35	46	7.4
Southeast 2		2.7	151	96	8.6
Midwest 2		2.9	99	N/A	9.3
West Coast		2.6	20	20	9.1

Table 5.6

Correlation coefficients relating water quality features with satisfaction and chlorine sensitivity

Water quality characteristic	Population	Satisfaction with... Overall quality	Taste	Appearance	Smell	Safety	Healthiness	Chlorine sensitivity
Disinfectant residual (at WTP)								
	FC markets	0.02	0.02	0.00	0.00	0.01	0.04	0.06
	CC markets	0.04	0.04	0.01	0.09	0.05	**0.13**	0.09
	FC & CC markets	0.01	0.02	0.00	0.00	0.02	0.04	---
Treated water temperature								
	FC markets	0.02	0.02	0.01	0.00	0.01	0.05	0.00
	CC markets	0.06	0.06	0.02	0.10	0.06	0.16	0.09
	FC & CC markets	0.08	0.08	0.02	0.01	0.02	0.08	---
TDS								
	FC markets	0.00	0.02	0.02	0.01	0.00	0.01	0.01
	CC markets	0.01	0.04	0.01	0.01	0.00	0.00	0.03
	FC & CC markets	0.02	**0.10**	0.00	0.00	0.00	0.00	---
Hardness								
	FC markets	0.00	0.00	0.00	0.02	0.00	0.01	0.03
	CC markets	0.00	0.01	0.00	0.01	0.00	0.00	0.04
	FC & CC markets	0.00	0.01	0.00	0.00	0.00	0.01	---

FC- Free chlorine
CC- Combined chlorine

In summary, there was no correlation between tap water hardness, alkalinity, or free chlorine residual concentration with threshold sensitivity to either free chlorine or monochloramine. In fact, the presence of a strong sulfurous odor in the Southeast 1 market (due to high hydrogen sulfide levels in their source water) was not often remarked upon in the survey and did not appear to mask the flavor of free chlorine as compared to the other free chlorine markets. The effect of background flavors is investigated further in Chapter 6.

Background Flavor Characterization of the Taste Test Source Waters

Samples of the finished waters used to prepare the taste test samples for each market were shipped overnight from each utility to the Ondeo Technology and Research Center. Some samples were destroyed in shipment (bottles broken), but the samples from the West, West Coast, Southeast 1, Midwest 1, Midwest 2, and East Coast were received intact and analyzed.

The samples were tested for the presence of odorous compounds by closed-loop stripping gas chromatography with mass spectrometry (CLSA-GC-MS), which is the method of choice for the earthy and musty descriptors. The objective of this analysis was to check whether specific samples could have contained high levels of odorants that could have partially masked the perception of chlorine by consumer panels. The most important compounds that were targeted were geosmin and 2-Methylisoborneol (2-MIB), as numerous authors have previously shown the possibility of masking effects between chlorine and earthy-musty descriptors.

Because of transportation time, the water samples were reassessed by flavor profile analysis soon after their arrival. Samples that still showed a chlorinous odor were dechlorinated with sodium thiosulfate prior to FPA, while samples submitted to CLSA were not dechlorinated. Four samples showed earthy-musty type tastes or odors:

- Midwest 1 - Intensity of 9 for the musty descriptor (on a scale of 1 to 12),
- Midwest 2 - Maximum taste and odor intensity of 10-12,
- West Coast - No odor, musty taste intensity of 6,
- Southeast 1 - Musty taste intensity of 6 to 8

Three samples showed the presence of geosmin (Midwest 1, Midwest 2 and West Coast). The Midwest 2 sample also contained 2-MIB. No earthy-musty odor metabolite could be detected in the Southeast 1 sample.

The concentrations of MIB and geosmin in the three above-mentioned samples were low, between 2 and 5 ng/L. This is in the range of the published odor threshold for geosmin and 2-MIB, hence there is an apparent discrepancy between the measured concentrations and the taste and odor intensities. Various cases of such a discrepancy have been observed at the Ondeo Technology and Research Center where samples with concentrations of geosmin-MIB close to the 1-2 ng/L limit of detection had odor intensities of 12 (Malleret 2001). The odor threshold of geosmin and MIB at 45°C was reassessed by Malleret (2001) who determined an odor threshold of 0.5 ng/l for the natural stereoisomer (-) geosmin. Therefore, it can be concluded that even the low levels of geosmin-MIB detected by CLSA-GC-MS could be responsible for the intense earthy-musty odors of the Midwest 1, Midwest 2, and West Coast samples. On the other hand, no explanation can be given for the musty taste of the Southeast 1 sample.

The East Coast water sample showed a low intensity phenolic-medicinal taste (intensity of 4 to 5), therefore this sample was extracted by methylene chloride liquid-liquid extraction (which is suited to the extraction of phenolic compounds or iodinated THM's that explain most of these types of tastes). No chlorophenols or bromophenols could be identified. The plastic additives T-butylphenol and 2,6-di-tertio-butylphenol were detected, together with an iodinated THM (dichloroiodomethane) and another iodo compound. These compounds likely explain the phenolic-medicinal taste of this sample.

The West market sample showed an intense swampy-septic odor that could not be explained due to the absence of dimethyldisulfide or dimethyltrisulfide in the CLSA extract. This odor might be an artifact of the dechlorination process with sodium thiosulfate applied prior to the FPA analysis. In fact, this sample was the only one that revealed the presence of phenol and chlorophenol. Their concentration could not be determined but was probably significant, as CLSA is known to very poorly extract phenol and chlorophenols. Their presence was unexpected, as this sample did not show any typical chlorophenol odor or taste, but a very strong bitter taste. A bitter aftertaste is often detected during the testing of solutions containing chlorophenols, hence the phenols in this sample may partially explain the bitter taste detected by the FPA panelists.

It is possible that the high average consumer threshold sensitivities measured in these markets (as compared to the literature values) were due to the masking effects of the musty/earthy, metallic and swampy odors detected by FPA analysis. However, as the panelists were not given multiple taste tests with the same water with different background flavor intensities, this hypothesis cannot be substantiated by the data collected.

Comparison of Taste Test Panelist Opinions with those of the Market Survey Respondents

Taste test (threshold sensitivity test) participants in all seven utility markets were less satisfied than market consumers (telephone survey) with respect to aesthetics and safety, except with respect to the question "I believe it is safe to drink water that has a chlorinous taste and odor" (Tables 5.7 and 5.8 and Figures 5.5 – 5.7). This trend was true for almost all market–question comparisons (40/42). The difference in average satisfaction between the market and taste test consumers was typically 5 to 15%.

Table 5.7
Average satisfaction with water quality with respect to aesthetics and safety in the free chlorine markets (Q005, Q010, Q015, Q020, Q025, Q030, Q212)

	East Coast		Southeast 1		Midwest 1		West		All	
	Market*	Taste test*	Market	Taste test	Market	Taste test	Market	Taste test	Market	Taste test
Number of samples	192	41	195	40	193	30	199	48	779	159
TW quality†	4.0	3.6	3.6	3.3	4.1	3.8	3.8	3.7	3.9	3.6
TW taste†	3.7	3.3	3.5	3.1	3.9	3.7	3.6	3.5	3.7	3.4
TW appearance†	4.6	4.0	4.2	3.8	4.4	4.0	4.3	3.7	4.4	3.9
TW smell†	4.3	3.4	4.0	3.6	4.4	3.9	4.1	3.7	4.2	3.6
TW safety††	4.4	4.2	4.1	4.0	4.4	4.2	4.4	4.2	4.3	4.1
TW healthiness§	4.0	4.0	3.7	3.6	4.2	4.0	4.1	3.8	4.0	3.8
Safe to drink water with chlorinous taste/odor**	2.7	3.5	2.7	3.1	2.8	3.3	3.1	3.1	2.8	3.3

* Market - respondents to telephone survey, taste test - panelists in chlorinous flavor threshold sensitivity tests

† 1. Very dissatisfied, 2. Dissatisfied, 3. Neither dissatisfied nor satisfied, 4. Satisfied, 5. Very Satisfied

†† 1. Unsafe, 2. Somewhat unsafe, 3. Neither safe nor unsafe, 4. Somewhat safe, 5. Safe

§ 1. Unhealthy, 2. Somewhat unhealthy, 3. Neither healthy nor un healthy, 4. Somewhat healthy, 5. Healthy

** 1. Disagree, 2. Somewhat disagree, 3. Neither agree nor disagree, 4. Somewhat agree, 5. Agree.

Table 5.8
Satisfaction with water quality with respect to aesthetics and safety in the combined chlorine markets (Q005, Q010, Q015, Q020, Q025, Q030, Q212)

	Southeast 2		West Coast		All	
	Market*	Taste test*	Market	Taste test	Market	Taste test
Number of samples	195	33	198	30	393	63
TW quality†	3.8	3.7	4.0	3.7	3.9	3.7
TW taste†	3.6	3.5	3.9	3.6	3.7	3.5
TW appearance†	4.1	3.5	4.3	4.0	4.2	3.7
TW smell†	4.2	3.8	4.4	3.8	4.3	3.8
TW safety††	4.3	3.8	4.3	3.9	4.3	3.8
TW healthiness§	4.1	3.3	4.2	3.6	4.1	3.5
Safe to drink water with chlorinous taste/odor**	2.6	2.7	2.8	2.8	2.7	2.8

* Market - respondents to telephone survey, Taste test - panelists in chlorinous flavor threshold sensitivity tests
† 1. Very dissatisfied, 2. Dissatisfied, 3. Neither dissatisfied nor satisfied, 4. Satisfied, 5. Very Satisfied
†† 1. Unsafe, 2. Somewhat unsafe, 3. Neither safe nor unsafe, 4. Somewhat safe, 5. Safe
§ 1. Unhealthy, 2. Somewhat unhealthy, 3. Neither healthy nor un healthy, 4. Somewhat healthy, 5. Healthy
** 1. Disagree, 2. Somewhat disagree, 3. Neither agree nor disagree, 4. Somewhat agree, 5. Agree.

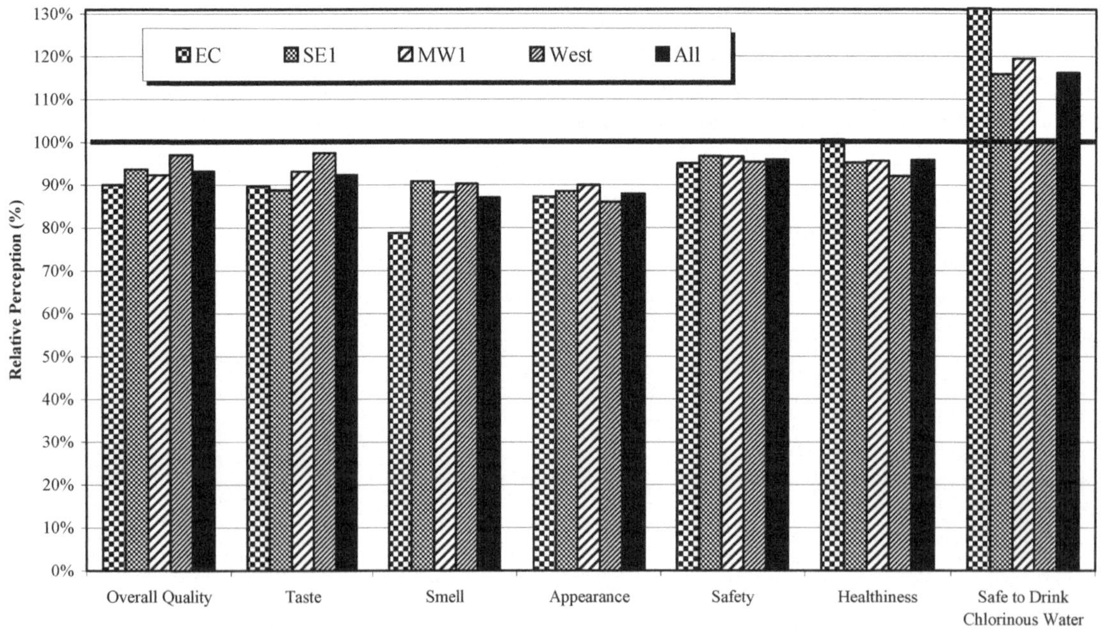

Figure 5.5 Comparison of average perceptions of the free chlorine target markets and taste test participants with respect to tap water aesthetics and safety - 100 % denotes a perfect match between the average market value and the average taste test value

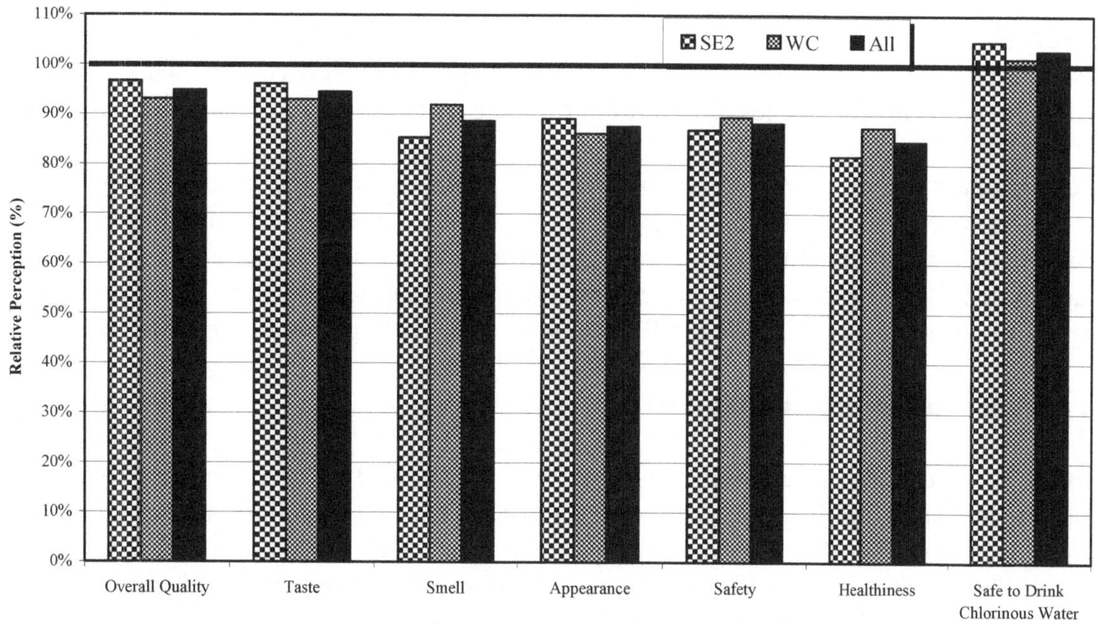

Figure 5.6 Comparison of average perceptions of the combined chlorine target markets and taste test participants - 100% denotes a perfect match between the average market and taste test values

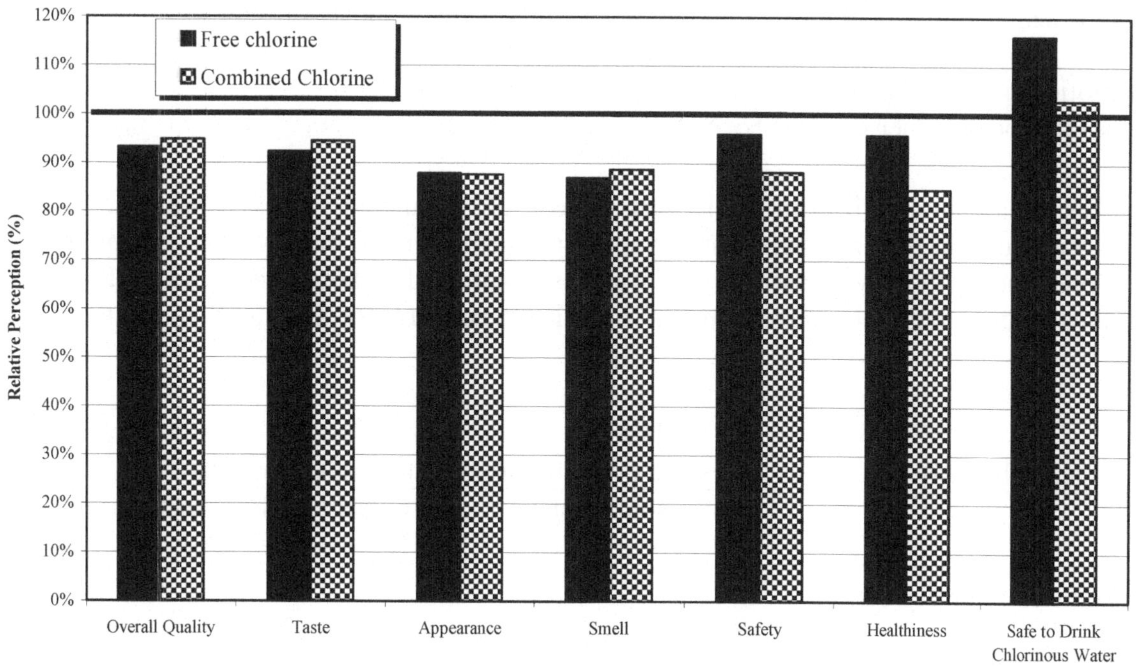

Figure 5.7 Comparison of average perceptions of the total free and combined chlorine target market and taste test participants - 100% denotes a perfect match between the average market and taste test values

These data indicate two things: 1) the taste test average consumer focus groups tended to attract individuals that were less satisfied than the average consumer in that market - focus group results may indicate less satisfaction than is actually present; 2) individuals who perceive chlorinous flavors to be unsafe are less likely to participate in chlorinous flavor sensitivity taste tests.

When comparing the answers given by the taste test groups to the remaining surveyed consumers, it is apparent that the taste test group underestimated average consumer satisfaction with water aesthetics and overestimated tap water safety associated with chlorinous flavors. This should be accounted for prior to drawing conclusions regarding tap water safety based on only a specific group of consumers (in this case in all those who were willing to undergo taste tests).

Taste Test Panelist Survey of Acceptability

A brief (3 question) acceptability survey was added at two free chlorine and one monochloramine test site (Southwest 1, Midwest 1 and Southeast 2) to solicit customer opinion about the appropriate level of free chlorine in tap water with respect to the two main routes of direct consumer interaction (and the source of most complaints with respect to chlorinous tastes and odors): drinking it and showering with it.

The questions posed were:

1. How much chlorine do you prefer in your water?
   a. No chlorine (meaning below your taste threshold),
   b. A little chlorine (enough so that you just barely taste it)
   c. A significant amount of chlorine (enough so that it is very obviously present)

2. Do you notice any chlorine odors when you shower? YES / NO
   If so, does this odor bother you? YES / NO

The results are summarized in Table 5.9. Almost all consumers surveyed prefer a low level of chlorine, but that they are split fairly evenly on whether or not they want to be able to taste it (and presumably thereby be assured that water is disinfected). These results also indicate that chlorinous odors in the shower are not a significant concern for the large majority of individuals.

Table 5.9
Acceptability of chlorinous tastes and odors reported by taste test panelists

	Southeast 1 (%)	Midwest 1 (%)	Southeast 2 (%)
Chlorine level preferred			
Below taste threshold	40	50	58
At the taste threshold	60	50	29
Significantly above the taste threshold	0%	0	13

(continued)

Table 5.9 (continued)

	Southeast 1 (%)	Midwest 1 (%)	Southeast 2 (%)
Chlorinous odors in the shower			
Don't notice it	68	87	87
Notice it	33	13	13
And find it objectionable	10	7	6

Differences between Consumer Groups

To better understand and characterize the impact of chlorine sensitivity on consumer satisfaction, a series of statistical tests were conducted to answer the question "Does the level of chlorine sensitivity affect a consumer's perception of tap water?" Because consumer sensitivity to chlorine was dependent on the type of chlorine residual used in a particular market, consumers were grouped into either the free or combined chlorine group. Consumer perception was addressed in four areas: satisfaction with overall quality, taste, appearance, and smell.

Within free and combined chlorine markets, there were no significant differences (i.e., p-values were all greater than 0.05) between consumers of different sensitivity levels and their perception of tap water characteristics (Table 5.10). This was confirmed by correlation tests. Each consumer's satisfaction level, scaled from 1 to 5, was plotted against chlorine sensitivity. Within free and combined chlorine markets, there was no significant correlation. This was expected based on the taste test results discussed previously.

Chlorine sensitivity by itself did not explain the differences in consumer perception of tap water. So the question remains: What factors affect consumer sensitivity to chlorine? Do markets with greater levels of chlorine residuals also have those consumers with lower sensitivities? Data show that statistical differences between markets did exist, but there was no statistically significant correlation (i.e., both R^2- values were less than 0.5).

Age and smoking habits were also examined in regards to chlorine sensitivity levels but found to be insignificant. Differences in satisfaction levels for each market were insignificant. Using chlorine sensitivity to categorize consumers' satisfaction with various water quality features was examined per respondent. Regardless of the market or water quality feature, no significant difference was found between consumers of different sensitivity levels. Initially, it had been hypothesized that consumers with lower levels of sensitivity might be more dissatisfied with various aspects of water quality, especially taste and smell, but this was not the case. Furthermore, correlations between satisfaction and threshold were also very poor.

The data presented in this chapter indicated that consumer sensitivity to chlorine is dependent on multiple factors and chlorine residual is only minor contributor to the issue of consumer sensitivity.

Table 5.10

Level of significance between groups of different chlorine sensitivity and related water quality features

Population	Correlation / Difference between…		Degree of correlation (R^2)*	Level at which significant difference exists between markets within each "population" (p-value)*
	Independent variable	Dependent variable		
FC markets†	Satisfaction w/ overall quality (Q005)	Cl_2 Sensitivity	0.02	0.63
FC markets	Satisfaction w/taste (Q010)	Cl_2 Sensitivity	0.04	0.18
FC markets	Satisfaction w/appearance (Q015)	Cl_2 Sensitivity	0.02	0.47
FC markets	Satisfaction w/smell (Q020)	Cl_2 Sensitivity	0.01	0.78
CC markets‡	Satisfaction w/ overall quality (Q005)	Cl_2 Sensitivity	0.04	0.33
CC markets	Satisfaction w/taste (Q010)	Cl_2 Sensitivity	0.04	0.31
CC markets	Satisfaction w/appearance (Q015)	Cl_2 Sensitivity	0.02	0.72
CC markets	Satisfaction w/smell (Q020)	Cl_2 Sensitivity	0.03	0.38
FC markets	Cl_2 Sensitivity	Cl_2 Residual	0.06	0.03
CC markets	Cl_2 Sensitivity	Cl_2 Residual	0.09	0.02

* $R^2 \geq 0.5$ and p-values ≤ 0.05 indicate significant correlations between two variables.

† FC - free chlorine

‡ CC - combined chlorine

CHAPTER 6
SENSITIVITY TO CHLORINOUS FLAVORS – COMPARISON BETWEEN TRAINED AND UNTRAINED PANELISTS

INTRODUCTION

As discussed in Chapter 2 and confirmed by this study, perception of chlorinous flavor (as opposed to no flavor or a fresh flavor) correlates to a reduced level of satisfaction with both tap water's aesthetics and healthiness/safety. Chlorinous flavors are a cause of concern for consumers and hence for water utility managers.

Chapter 5 reported the results of a seven-market study on consumer sensitivity to chlorine in consumers' own drinking waters. While assessing customer sensitivity to chlorine in their own tap water is very valuable, it does have some limitations. Natural waters are complex and have a variety of background flavors that can influence consumer sensitivity to chlorine and consumer identification of the objectionable tastes or odors (e.g., minerals or geosmin).

In addition, "average" consumers may often misidentify the flavors they detect in their tap water, resulting in erroneous reporting of taste and odor problems. The use of trained panelists via FPA allows more accurate, repeatable characterization of flavors and flavor intensities. FPA can be a helpful option for water utilities wishing to better identify taste and odor problems in their system. Usually, these groups include five panelists (instead of 10 untrained panelists) and can identify flavor and/or odors quite accurately. This is extremely advantageous as FPA panels can provide utility personnel with enough information to identify the source of a taste and odor problem. There is one significant drawback to FPA panels: the use of trained panelists may also lead to erroneous conclusions about how to solve a taste and odor problem as the sensitivity and accuracy of a FPA panel may be different from most of the average consumers.

Since plant personnel have to deal with the average consumer satisfaction or dissatisfaction, it was important to investigate how the sensitivity of trained panelists compare to that of untrained panelists.

OBJECTIVES

The objective of the work presented in this chapter was threefold: to compare expert (FPA) panelist results to average consumers, to compare the sensitivity of different consumer groups (France versus the United States, typical residuals of 0.1 mg/L Cl_2 versus 1 mg/L Cl_2) to the same water-chlorine combinations and to investigate the effects of some background flavors on consumer sensitivity.

The results of this work are presented in three sections:

1. Measure and compare the threshold sensitivity of an FPA panel and average American and French consumers to different oxidants used in drinking water disinfection (e.g., chlorine, monochloramine and chlorine dioxide) using the same test water (Evian[1]).
2. Test the ability of trained and untrained panels to correctly identify background flavors.

[1] Evian™ spring water, Evian-les-Bains, France.

3. Investigate the antagonism/synergism effects from background hardness and water temperature.

The complete results of the human sensitivity tests presented in this Chapter are detailed in Appendix E.

METHODOLOGY

Identification of drinking water disinfectant thresholds was performed by expert (trained) panels according to the FPA protocol APHA Method 2710. Threshold sensitivity evaluations were completed according to the ASTM Method E679-91 (Forced Choice Triangle Test). Comparison of sensitivities between trained (FPA) and untrained (average consumer) panelists was performed according to the Forced Choice Triangle Test, and in some cases, also using a presence/absence test. A presence/absence test was used to investigate whether chlorinous flavors were correctly identified by average consumers. It was also used to measure the impact of background water quality parameters (total dissolved solids, musty flavors, and temperature) on the perception of chlorinous flavors.

Sensitivity of Trained and Untrained Panelists to Different Chlorinous Flavors

Trained Panelist Tests

The FPA procedures were performed by four to five panelists. Intensity was calculated as the median of each panelist answer (flavor perception) for each disinfectant concentration. A reference scale, composed of three concentrations of sugar (sucrose), was used to calibrate the FPA panelist flavor intensity scale (5 g/L, 10 g/L, and 30g/L, respectively, representing 4 [weak], 6 [weak to moderate], and 10 [moderate to strong]). The tests were carried out at 25°C. Results are presented in Weber-Fechner plots, where intensity is assumed to be a linear function of log of the chlorine concentration:

$$\text{Intensity} = C * \text{Log [Chlorine Concentration in mg/L as } Cl_2] \quad (1)$$

Taste detection limits were deduced from the Weber-Fechner graphs to an intensity of 1, which equals the threshold taste.

The Forced Choice ASTM procedure was performed by four to five panelists according to the method previously described. Different concentrations of chlorine (0.05 to 0.5 mg/L), monochloramine (0.1 to 1.0 mg/L), and chlorine dioxide (0.1 to 1.5 mg/L as Cl_2O_2) were prepared in Evian water. Samples were brought to 25°C, at which time each disinfectant concentration sample and two taste-free reference samples (Evian) were presented to the panel. For each taste testing panel session, a minimum of five different concentrations were presented to the panelists from lowest to highest disinfectant concentration to avoid panelist fatigue. The pH of the Evian water was 7.5 and did not vary after chemical addition.

Untrained Panelist Tests

The untrained panel was composed of 10 tasters, and followed the presence/absence method (P/A) and/or the ASTM triangle test. The forced triangle test was performed as described for the trained panelists. In the P/A procedure, one sample containing the target compound was presented to each panelist, which had to report whether or not he/she perceived a chlorinous flavor (Yes/No). The percentage of positive answers was plotted as a function of concentrations expressed in Log units, in a Weber-Fechner representation:

$$\% \text{ Positive Answers} = C * \text{Log [Chlorine Concentration in mg/L as } Cl_2] \qquad (2)$$

The flavor threshold was determined from the graph for a 50% positive answer value reported by the panel. Different concentrations of chlorine (0.1 to 1.0 mg/L), monochloramine (0.1 to 1.0 mg/L), and chlorine dioxide (0.1 to 1.2 mg/L ClO_2), were prepared in Evian water. For each taste panel session, a minimum of five different concentrations were presented to the panelists from the lowest to the highest disinfectant concentration to avoid sensory fatigue.

Recognition of Chlorinous Flavors by Untrained Panelists

The objective of these experiments was to demonstrate that "chlorinous" flavors were not well recognized by average consumers (untrained panelists). In the first experiment, performed at 20°C, taste-free water (Evian) samples containing disinfection by-products (DBPs) or chlorine at concentrations greater than the measured threshold (Table 6.1) were given to untrained panelists in France; the panelists were asked to describe the flavor of the samples.

Table 6.1
Concentrations of the taste test solutions

Compounds	Concentration
Free chlorine	0.15 mg/L
Iodoform	5 µg/L
2,4,6, Tribromophenol	10 µg/L
Pentachlorophenol	80 µg/L

A second experiment, also conducted at 20°C, included a comparison between samples containing free chlorine and one sample containing iodoform at concentrations above the threshold concentration. Panelists had to choose which sample tasted "different".

Impact of Background Water Quality on the Sensitivity of Trained and Untrained Panelists to Chlorinous Tastes

Influence of Water Temperature

To investigate the influence of water temperature on the perception of free chlorine and monochloramine, chlorine solutions were prepared in taste-free water (Evian) and incubated at 15, 20, and 25°C, respectively. Flavor analyses were performed on these samples according to the Forced-Choice taste test procedure with ten untrained panelists, and according to the FPA protocol with five trained panelists.

Influence of Background Minerals

Three mineral waters (Volvic[2], Evian, and Contrex[3]) of varying hardness and TDS content (Table 6.2) were chosen as reference (taste-free) waters for chlorinous flavor sensitivity threshold tests. Five samples of increasing concentrations (0.1 to 1 mg/L Cl_2) were given to the average French and American panelists, where only chlorine descriptor answers were evaluated.

Table 6.2
Main water quality characteristics of the three tested water

Parameter	Evian	Volvic	Contrex
pH (20°C)	7.5 - 7.6	7.2 - 7.3	7.3 - 7.4
$KMnO_4$ oxidability (mg/L O_2)	0.25	0.3	0.15
TDS 105°C (mg/L)	370	130	1400
Carbon dioxide (mg/L CO_2)	15	6	28
Alkalinity (mg/L $CaCO_3$)	298	46	1498
Conductivity (20°C -υS/cm)	505	140	1890

Influence of a Background Earthy/Musty Taste

The antagonistic effect of a musty taste on chlorine perception was first investigated with a solution of drinking water containing 0.20 - 0.25 ng/L of 2,4,6-trichloroanisole that has a Flavor Threshold Number (FTN) of 3. Then, background musty-taste solutions were prepared by adding 2,3,6-thrichloroanisole (2,3,6-TCA) to the taste-free water. Three solutions with an increasing FTN were prepared (Table 6.3).

Table 6.3
Characteristics of the musty taste background solutions

Solution	2,3,6-TCA concentration	FTN
#1	35 ng/L	5
#2	75 ng/L	10
#3	140 ng/L	20

Increasing concentrations of chlorine (0.1 - 1.0 mg/L) were added to these solutions. The diluted solutions were then submitted to four trained panelists to characterize chlorine perception (Chlorine Flavor Threshold). A P/A method was also applied (i.e., panelists had to determine if they perceived chlorine or musty taste for the different solution).

[2] Volvic mineral water, Auvergne, France
[3] Contrex mineral water, Contrexéville, France

RESULTS AND DISCUSSION

Sensitivity of Trained and Untrained Panelists to Different Chlorinous Flavors

Trained Panelists Tests

Free chlorine threshold. Figure 6.1 is an example of a Weber-Fechner curve, and summarizes the results obtained for the FPA flavor intensity test. A free chlorine flavor threshold (intensity of 1) value of 0.064 mg/L Cl_2 was measured. The minimum and maximum values were 0.037 and 0.092 mg/L Cl_2, respectively, at a 95% confidence level.

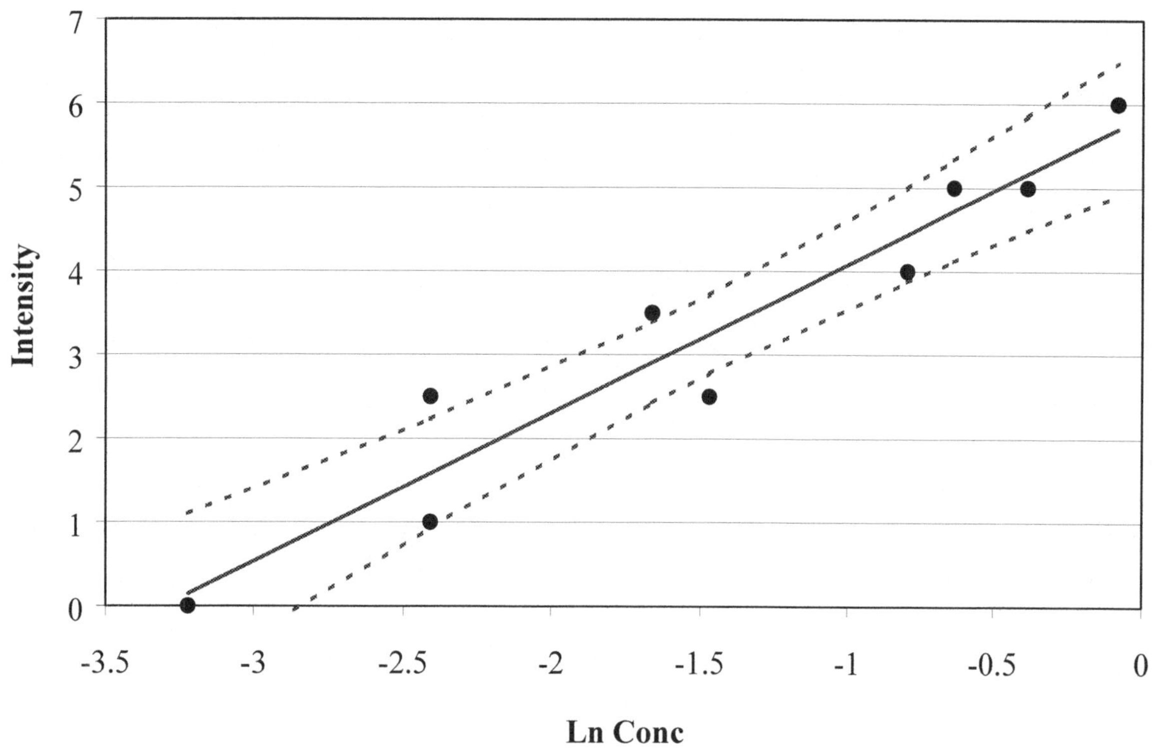

Figure 6.1 Weber-Fechner curve obtained by trained panelists for free chlorine

The results of the two experiments performed according to the Forced-Choice Triangle Test are summarized in Table 6.4. Figure 6.2 shows a plot of the percentage of positive answers (detection of free chlorine at a given concentration) versus the free chlorine concentration expressed in natural logarithmic units. The free chlorine threshold for the group was 0.054 mg/L Cl_2. The minimum and maximum values measured were 0.044 and 0.066 mg/L Cl_2, respectively, at a 95% confidence level.

Table 6.4
Free chlorine flavor threshold determined by the Forced-Choice procedure

Test Date	Threshold taste (mg/L Cl_2)	Std deviation (mg/L Cl_2)
04/06/01	0.05	0.03
11/22/01	0.04	0.011

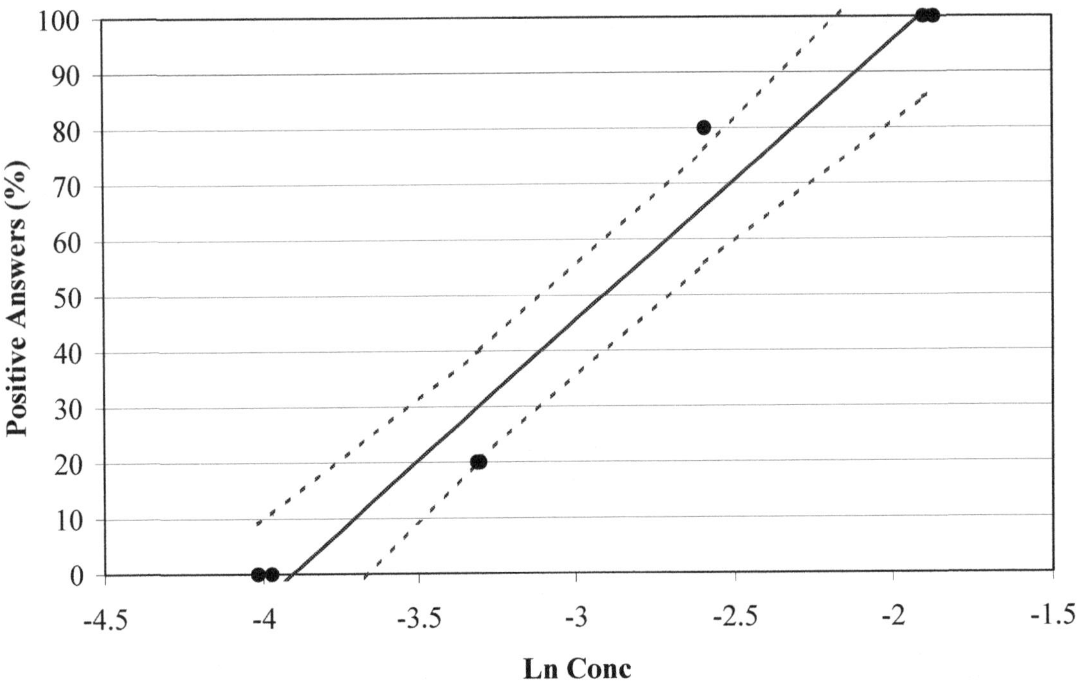

Figure 6.2 Panelist threshold sensitivity to free chlorine as measured by the Forced-Choice Triangle test

In these experiments, the free chlorine flavor threshold measured was 0.05 mg/L Cl_2 ± 50%. This value is significantly lower than those reported by Krasner & Barrett 1984 (Odor threshold reported = 0.24 to 0.30 mg/L Cl_2). Bryan et al. (1973) performed triangular taste testing at 40°C and reported a flavor threshold of 0.05 mg/L Cl_2 in the case of a non pH-adjusted chlorine sample.

Chloramine threshold. The same experimental approach used in assessing sensitivity to free chlorine was used to assess the panelists' perception of chloramines. All graphics and procedures are included in Appendix E. A monochloramine flavor threshold value of 0.15 mg/L Cl_2 was measured using the P/A test. The minimum and maximum values measured were 0.11 and 0.18 mg/L Cl_2, respectively, at a 95% confidence level.

The two taste testing experiments performed according to the ASTM method are summarized in Table 6.5. A monochloramine flavor threshold concentration value of 0.09 mg/L Cl_2 was measured. The minimum and maximum values measured were 0.022 and 0.13 mg/L Cl_2, respectively, at a 95% confidence level. This threshold value is significantly lower than those reported by Krasner & Barrett in 1984 (0.48 mg/L Cl_2).

Table 6.5
Monochloramine flavor threshold by Forced-Choice procedure

Test date	Threshold ($mg/L\ Cl_2$)	Std deviation ($mg/L\ Cl_2$)
11/28/01	0.07	0.024
05/03/01	0.11	0.095

As expected, the panelists had a higher threshold sensitivity to chloramine than to free chlorine. Monochloramine induces some sensation (metallic) in addition to the taste at low concentration.

Chlorine dioxide threshold. A chlorine dioxide flavor threshold concentration value of 0.2 mg/L Cl_2 was measured (minimum and maximum values measured were 0.13 and 0.27 mg/L Cl_2, respectively, at a 95% confidence level) via a P/A test.

Table 6.6 shows the chlorine dioxide flavor threshold measured via the Forced-Choice Triangle Test. A flavor threshold value of 0.19 mg/L ClO_2 ± 50% was measured for chlorine dioxide. The minimum and maximum were 0.15 and 0.21 mg/L ClO_2, respectively, at the 95% confidence level. This is similar to the values reported in the literature (0.20 to 0.25 mg/L, Hoehn et al. 1990).

Table 6.6
Chlorine dioxide flavor threshold by Forced-Choice procedure

Test date	Threshold ($mg/L\ ClO_2$)	Std deviation $mg/L\ CLO_2$
05/10/01	0.15	0.10
11/23/01	0.18	0.085

Untrained Panelists Tests

Free chlorine threshold - French panel. A free chlorine flavor threshold taste concentration value of 0.2 mg/L Cl_2 (Table 6.7) was measured. The minimum and maximum values measured were 0.15 and 0.27 mg/L Cl_2, respectively, at a 95% confidence level. As with the trained panelist tests, procedures and data can be found at the beginning of this chapter and in Appendix E, respectively.

Table 6.7
Free chlorine flavor threshold measured by Forced-Choice procedure - French panel

Test date	Threshold ($mg/L\ Cl_2$)	Std deviation ($mg/L\ Cl_2$)
04/10/01	0.29	0.48
08/09/00	0.2	0.42

Free chlorine - American panel. A free chlorine flavor threshold value of 0.93 mg/L Cl_2 (Table 6.8) was measured. The minimum and maximum values measured were 0.69 and 0.84 mg/L Cl_2, respectively, at the 95% confidence level.

Table 6.8
Free chlorine flavor threshold by Forced-Choice procedure - American panel

Test date	Threshold ($mg/L\ Cl_2$)	Std deviation $mg/L\ Cl_2$
08/01/01	0.89	0.58
07/31/01	0.76	0.56

Figure 6.3 compares the sensitivity of the untrained American panel and the trained panel to free chlorine. Although some untrained panelists had sensitivities as low as the trained panelists, on average they were significantly less sensitive. The free chlorine concentrations required to obtain positive answers from the majority of the untrained panelists were significantly higher than those required by the trained group.

This is in line with expectations—trained panelists as a group are almost always significantly more sensitive to flavors than untrained panelists as a group.

Figure 6.4 compares the sensitivity of the American and the French untrained panels to free chlorine. The same slope was found for both panels, but there was a definite shift in the flavor perception trend. The French panel was more sensitive to lower concentrations, but the increase of percentage of positive perception followed the same trend as the American panel. There was the same heterogeneity in the answers of both panels, likely because they were both untrained. The difference between panels is likely due to chlorination residual concentration practices in the two countries (0.1 - 0.2 mg/L Cl_2 in France versus 1.0 mg/L Cl_2 in the American). The French panel was likely less de-sensitized to free chlorine than the American panel.

Chloramine perception - French panel. A monochloramine flavor threshold value of 0.18 mg/L Cl_2 (Table 6.9) was measured. The minimum and maximum values measured were 0.13 and 0.23 mg/L Cl_2, respectively, at a 95% confidence interval.

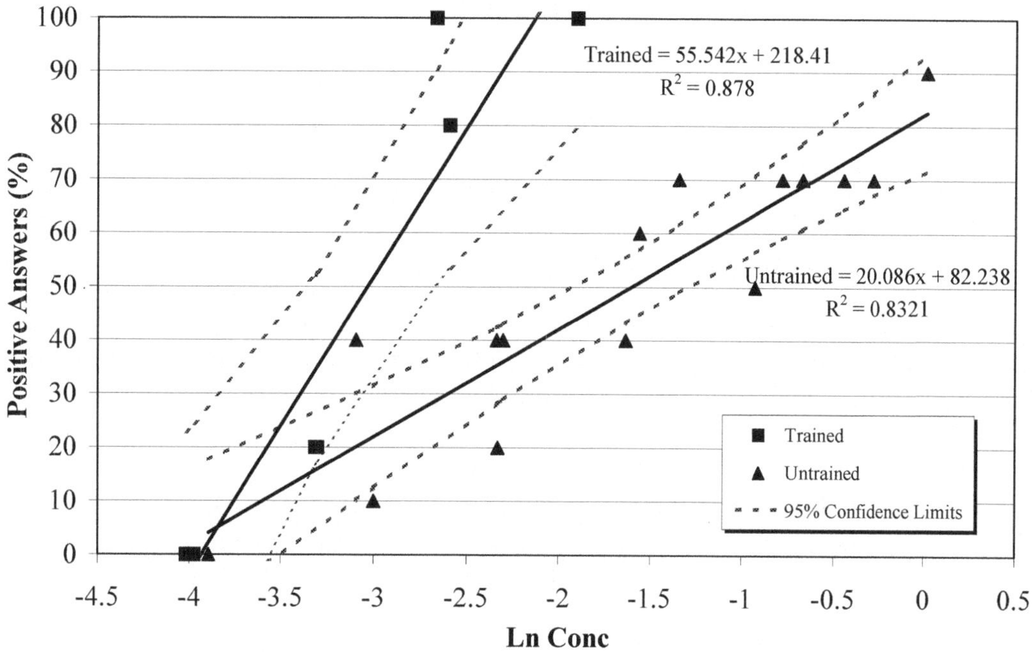

Figure 6.3 Comparison of chlorine perception for untrained American versus trained panel

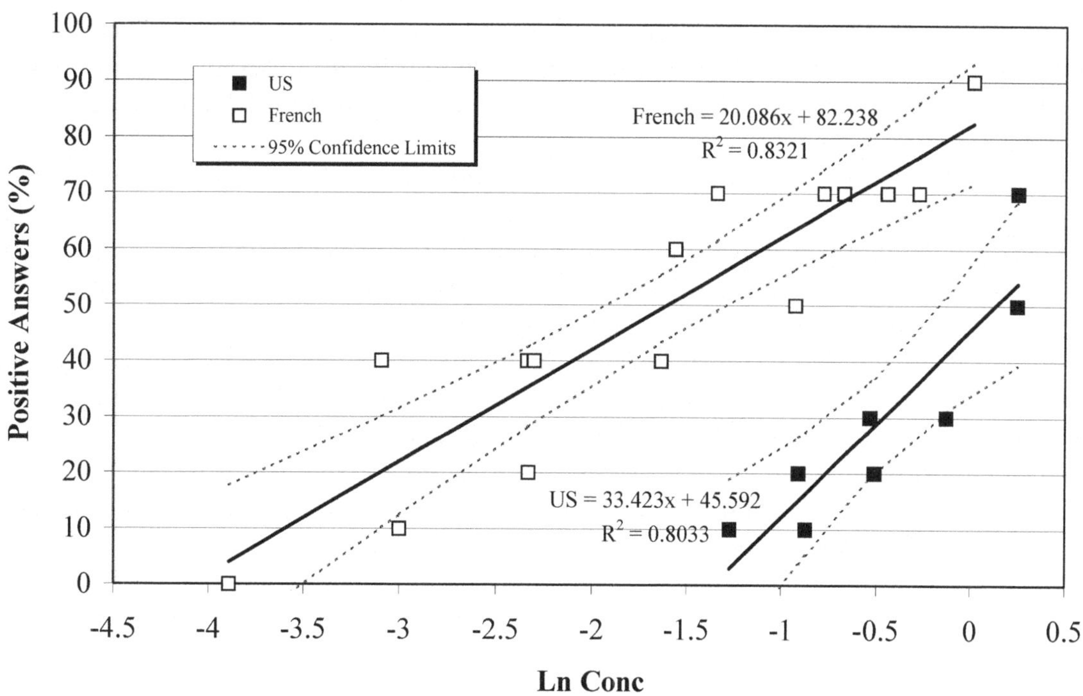

Figure 6.4 Comparison of untrained French and American panels' sensitivity to free chlorine

Table 6.9
Chloramine flavor threshold by Forced-Choice procedure - French panel

Test date	Threshold ($mg/L\ Cl_2$)	Std deviation ($mg/L\ Cl_2$)
04/19/01	0.22	0.58
08/22/00	0.27	0.39

Monochloramine perception - American panel. A monochloramine flavor threshold value of 1.8 mg/L Cl_2 (Table 6.10) was measured. The minimum and maximum values measured were 1.4 and 2.4 mg/L Cl_2, respectively, at a 95% confidence level.

Table 6.10
Chloramine flavor threshold by Forced-Choice procedure - American panel

Test date	Threshold ($mg/L\ Cl_2$)	Std Deviation ($mg/L\ Cl_2$)
02/08/01	1.22	1.32
02/08/01	0.85	0.95

Figure 6.5 compares the sensitivity of the American untrained panel and the trained panel to monochloramine. As with free chlorine, the main difference between the panels responses was the slope of the fit to the plot. Both untrained panels were significantly less sensitive to monochloramine than the trained panel. Also, because the group was untrained, their answers were more heterogeneous, leading to a higher chlorine concentration required to produce positive answers by the majority of the panelists.

Figure 6.6 compares the sensitivity of the American and French untrained panels to monochloramine. As observed with free chlorine, the same slope was found for both groups. The difference between the two panels was mainly a shift in the flavor perception. The French panel was more sensitive to lower concentrations of chloramine, but the change in percentage of positive perception with increasing concentration followed the same trend as the American panel. The same heterogeneity in the answers of both panels was observed because they were both untrained. As with the free chlorine thresholds, this difference in sensitivity is likely linked to the difference in chlorination residual concentration practices in the two countries.

Chlorine dioxide perception - French panel. A chlorine dioxide flavor threshold value of 0.17 mg/L ClO_2 was measured (Table 6.11). The minimum and maximum values measured were 0.085 and 0.26 mg/L ClO_2, respectively, at a 95% confidence level.

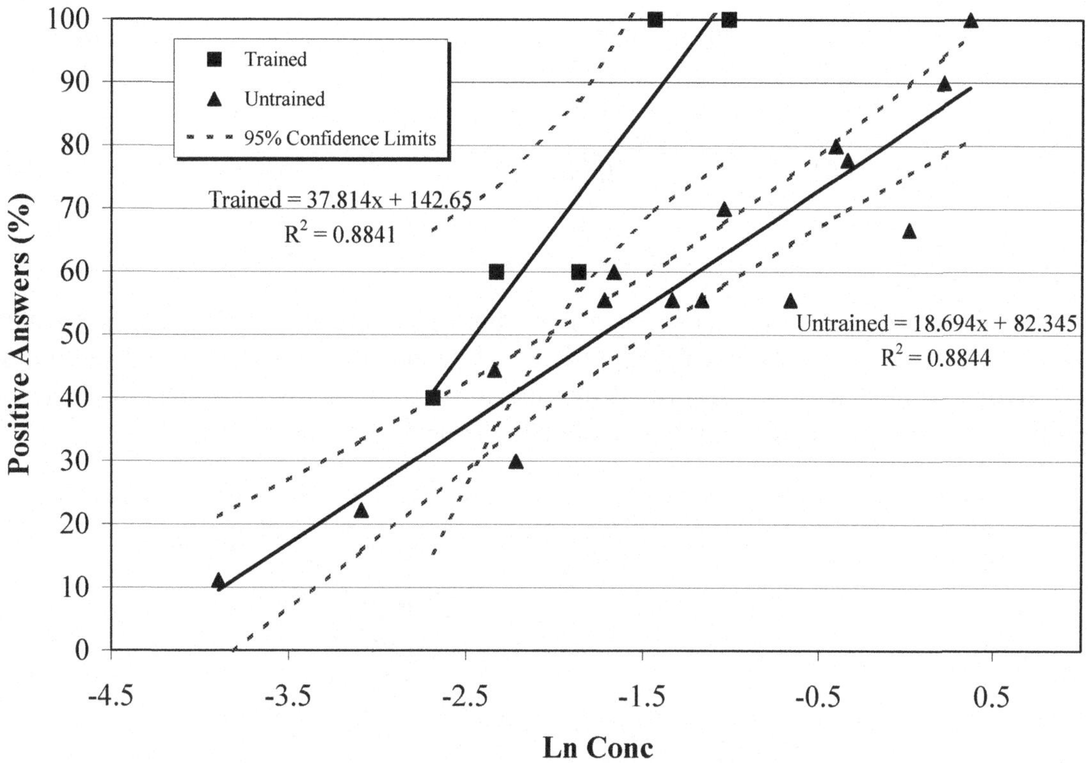

Figure 6.5 Comparison of monochloramine perception of the untrained American and the trained panel

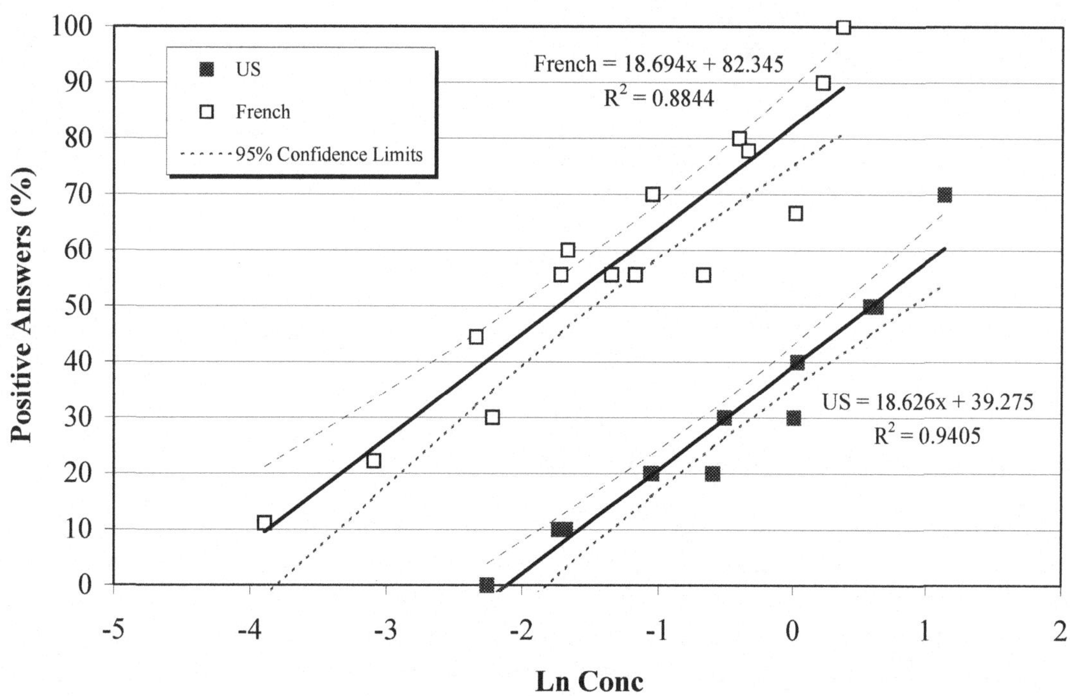

Figure 6.6 Comparison of monochloramine perception of the untrained French and U.S. panels

Table 6.11
Chlorine dioxide flavor threshold by Forced-Choice procedure - French panel

Test date	Threshold (geometric mean: mg/L ClO_2)	Std deviation mg/L ClO_2
05/10/01	0.08	0.07
04/24/01	0.16	0.125

Figure 6.7 compares the sensitivity of the untrained French and the trained panels to chlorine dioxide. There is no statistically significant difference between the trained and untrained panels. These results are noticeably different from results of the free chlorine and monochloramine tests, where differences between the panels were significant.

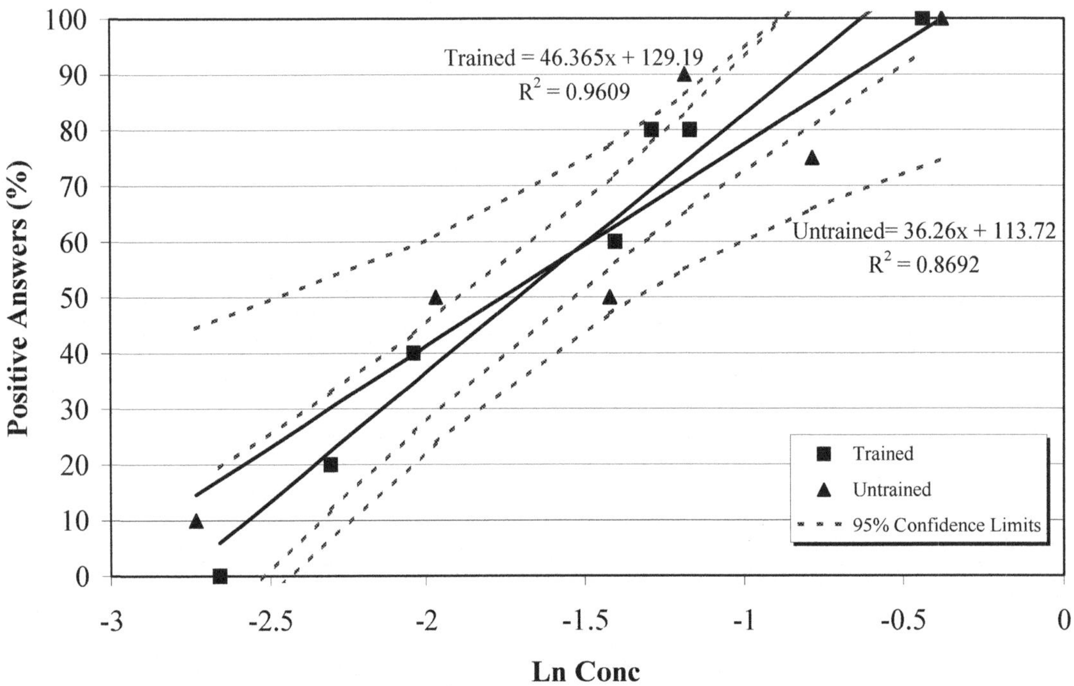

Figure 6.7 Comparison of chlorine dioxide perception of the untrained French and trained panels

Recognition of Chlorinous Flavors by Untrained Panelists

Untrained panelists were given samples that contained DBPs or free chlorine at concentrations greater than the threshold value. The panelists were asked to describe the flavor of each. The results are presented in Table 6.12. The different flavor descriptors given by the panelists were then organized into four main flavor categories: medicinal, chlorine-swimming pool, chemical, and other. These results indicated that untrained panelists are not able to determine the difference between chlorine, medicinal and chemical descriptors, (i.e., “average consumers” have a difficult time accurately describing flavors in water.)

Table 6.12
Flavor recognition test for chlorine and DBPs at 20°C

	Iodoform (5 μg/L) %	Chlorine (0.15 mg/L) %	2,4,6-Tribromo Phenol (10 μg/L) %	S4- Evian Blank %	Pentachloro Phenol (80 μg/L) %
Medicinal	45	18	27		9
Chlorine / swimming pool	45	73	18		9
Chemical	9		36		9
Other	9	18		27	73

A second experiment was performed using the ASTM Forced-Choice Triangle test to compare chlorine and medicinal iodoform tastes. All the untrained panelists recognized the difference between the iodoform and free chlorine tastes. However, most panelists could not accurately describe the iodoform taste (medicinal). Iodoform is a standard compound used to define the taste "medicinal" in FPA analysis. Other descriptors used were chemical/hydrocarbon, hydrocarbon/burned rubber, musty/pungent, chemical and chlorinous.

In conclusion, the two experiments show that untrained consumers were able to distinguish between different flavor types (chlorine, medicinal or chemical), but were not able to accurately or consistently (between group members) describe them. This indicates that, in many cases, it is likely difficult to accurately identify a consumer's taste and/or odor problem based solely on their description of that taste/smell.

Impact of Background Water Quality on the Sensitivity of Trained and Untrained Panelists to Chlorinous Tastes

Influence of Water Temperature

Untrained French panel. Figure 6.8 shows how temperature impacted the free chlorine flavor threshold of untrained French panelists. No significant differences in sensitivity were measured between 15 and 25°C. The same approach was taken and the same results were seen with monochloramine (Figure 6.9).

Trained panel. The same temperature experiments were performed with the trained panel and the results are shown in Figure 6.10 and 6.11. Although some differences in temperature were identified by the trained panelists, these were not statistically significant.

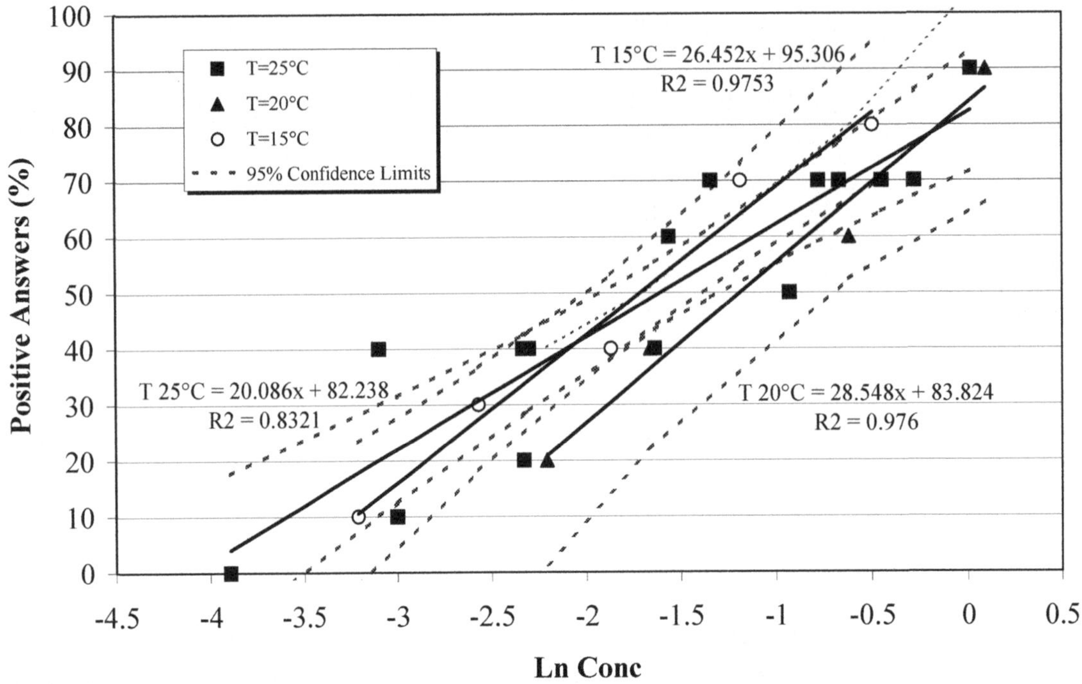

Figure 6.8 Impact of temperature on the perception of free chlorine by untrained French panelists

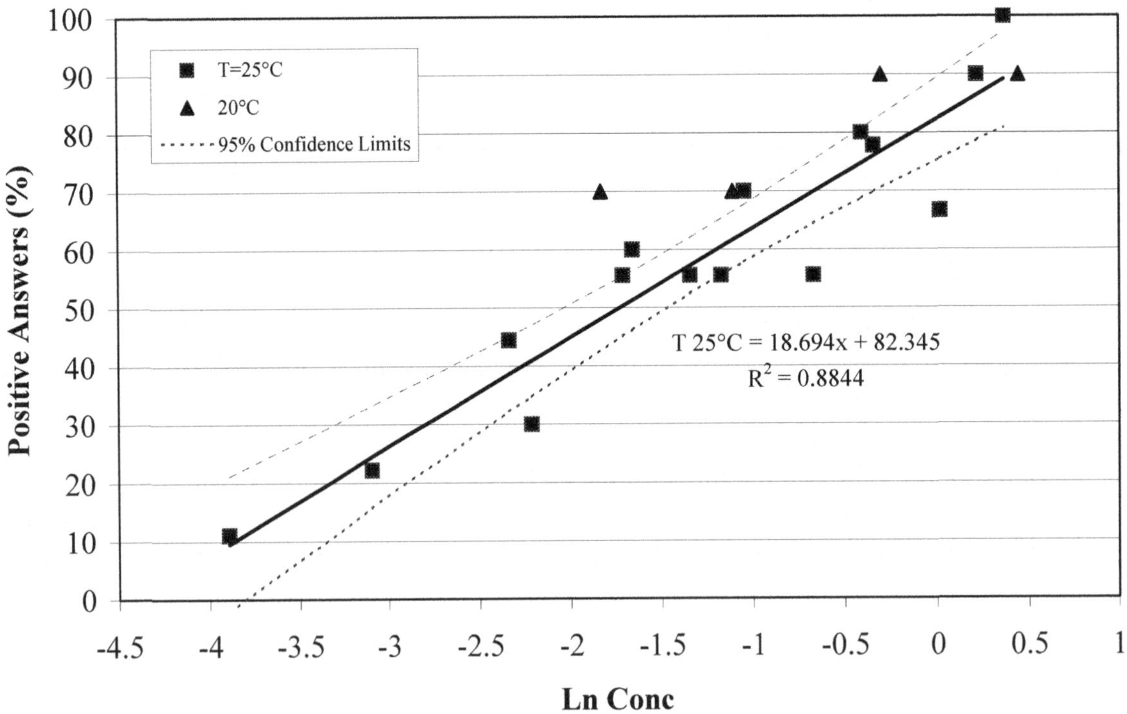

Figure 6.9 Impact of temperature on the perception of monochloramine by untrained French panelists

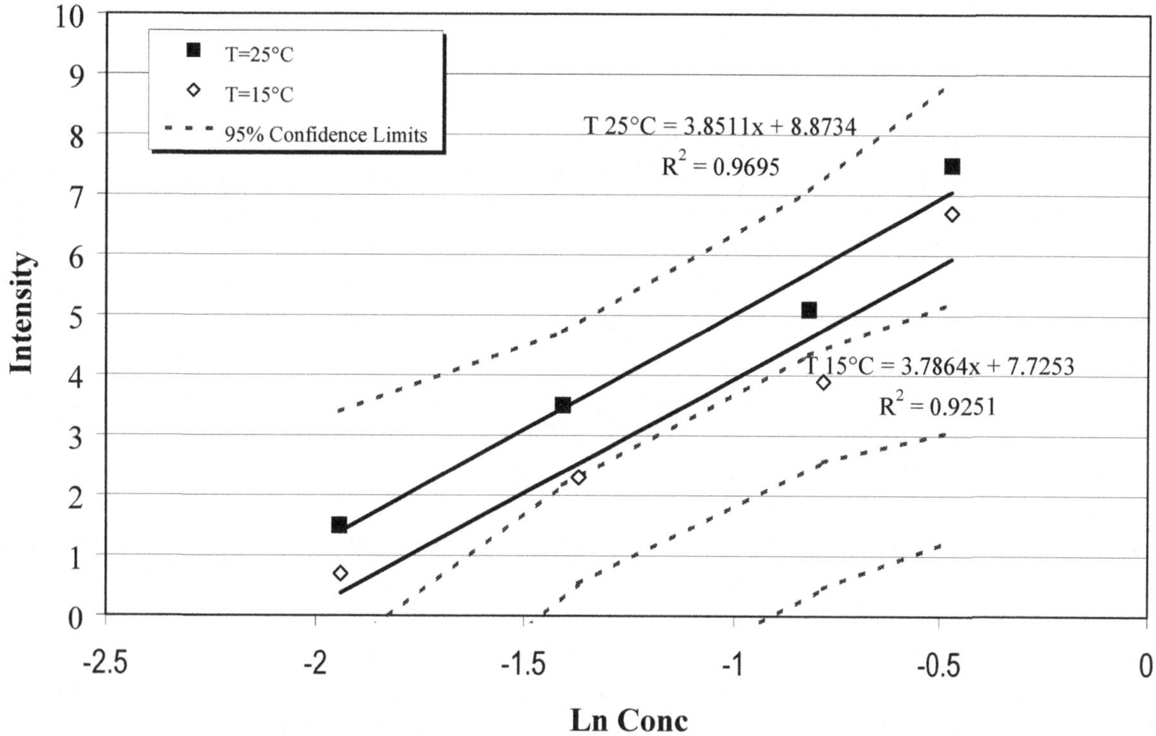

Figure 6.10 Impact of temperature on the perception of free chlorine by trained panelists

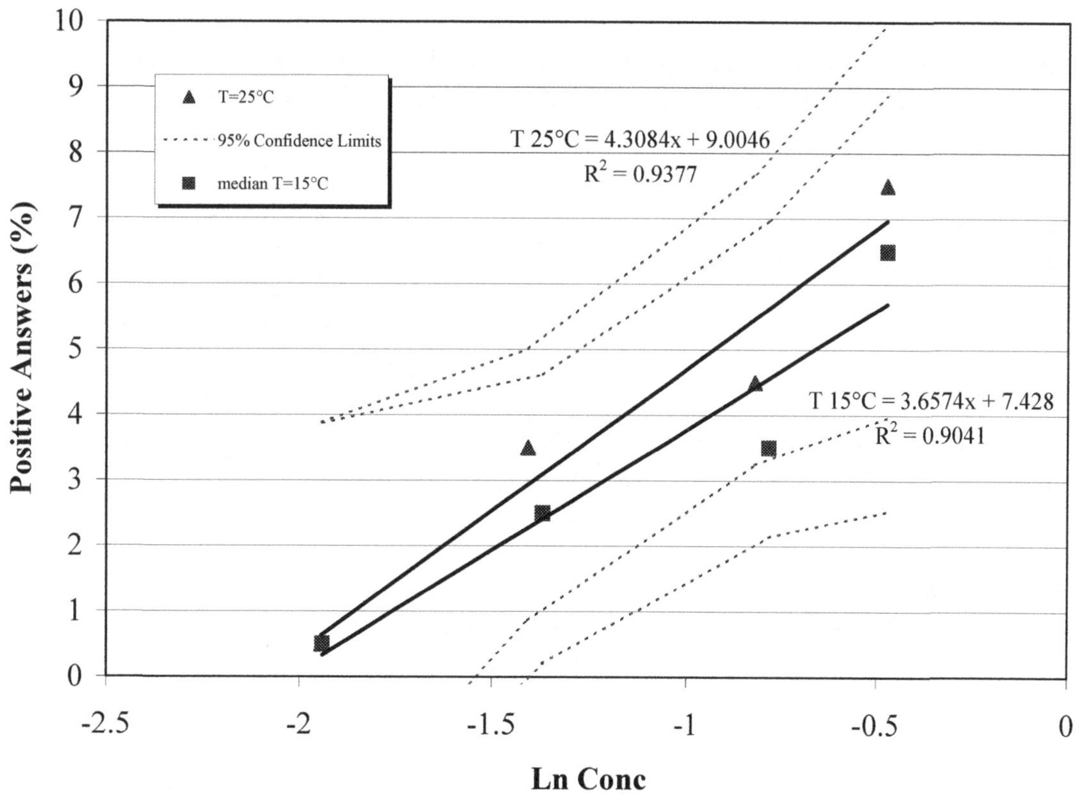

Figure 6.11 Impact of temperature on the perception of monochloramine by trained panelists

Forced-choice triangle tests were performed with free chlorine and chloramines using Contrex and Volvic bottled water and compared to those obtained using Evian water.

Untrained French panel. Figure 6.12 illustrates the impact of TDS on the perception of free chlorine by the untrained French panel. No statistically significant difference was detected (95% confidence level). Tests performed with chloramine are shown in Figure 6.13. They indicate that the sensitivity of the panelists to the chlorine flavor of chloramines was affected by TDS at low chloramine concentrations (0.1 to 0.3 mg/L Cl_2,) but not at higher concentrations (0.3 to 1.2 mg/L Cl_2).

Untrained American panel. Figure 6.14 shows the impact of TDS on the perception of free chlorine by the untrained American panel. Although no statistically significant influence of TDS was observed at a 95% confidence level, it was noticed that their sensitivity was affected by the type of water used. Lower thresholds were detected with Contrex than with the other waters. This is somewhat contrary to expectations—that the high background TDS would mask the chlorinous flavor.

Trained panel. The same experiments performed for the untrained panel were performed with the expert panel. Figures 6.15 and 6.16 show the influence of TDS on both average and median free chlorine perception. Although no statistically significant influence of TDS was noticed by the trained panelists at a 95% confidence level, the minimum chlorine thresholds observed were in the lower-TDS water (Volvic).

Some influence of water characteristics on chlorine or chloramine was noticed, but no clear links with TDS concentration was established. This result was also reported by Wiesenthal and Suffet (2002). They concluded that the effect of TDS content on chlorine and chloramine perception was more complex than expected because both antagonistic and synergistic effects of TDS (NaCl) may occur depending on the ratio of chlorine and TDS concentrations.

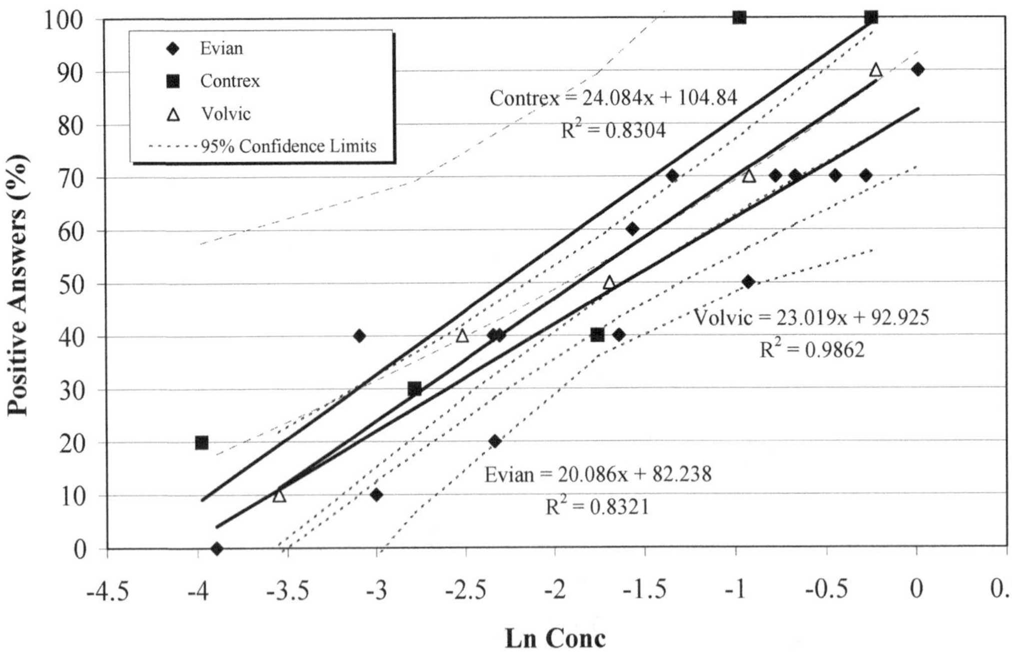

Figure 6.12 Impact of TDS on the perception of free chlorine by untrained French panelists

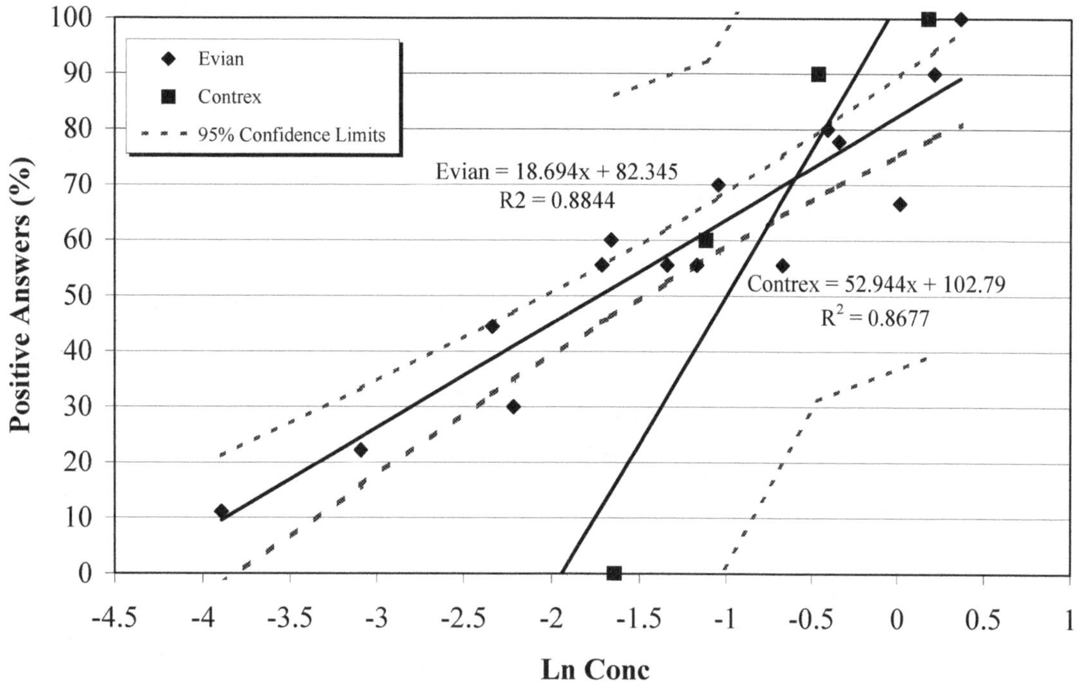

Figure 6.13 Impact of TDS on the perception of chloramine by untrained French panelists

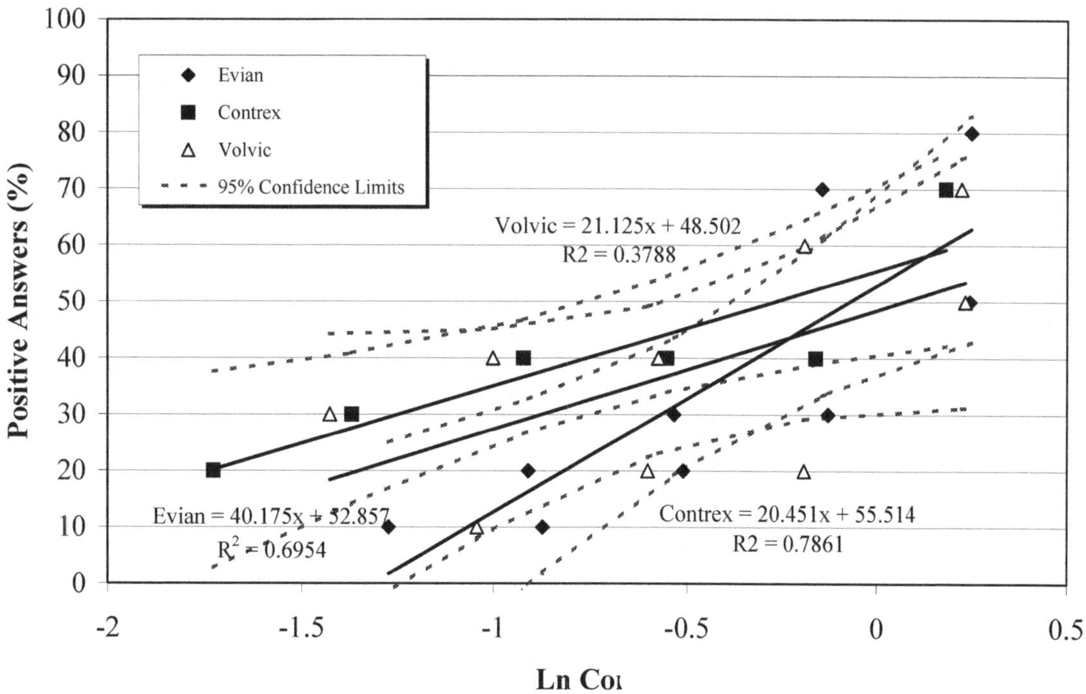

Figure 6.14 Impact of TDS on the perception of free chlorine by untrained American panelists

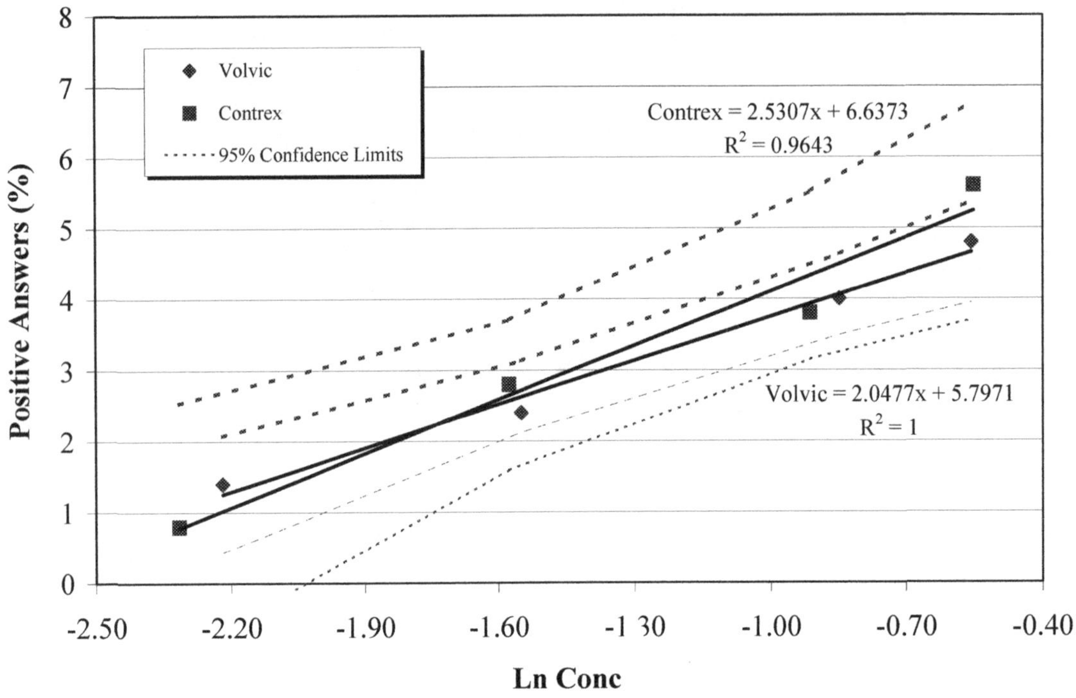

Figure 6.15 Impact of TDS on the average perception of free chlorine by trained panelists

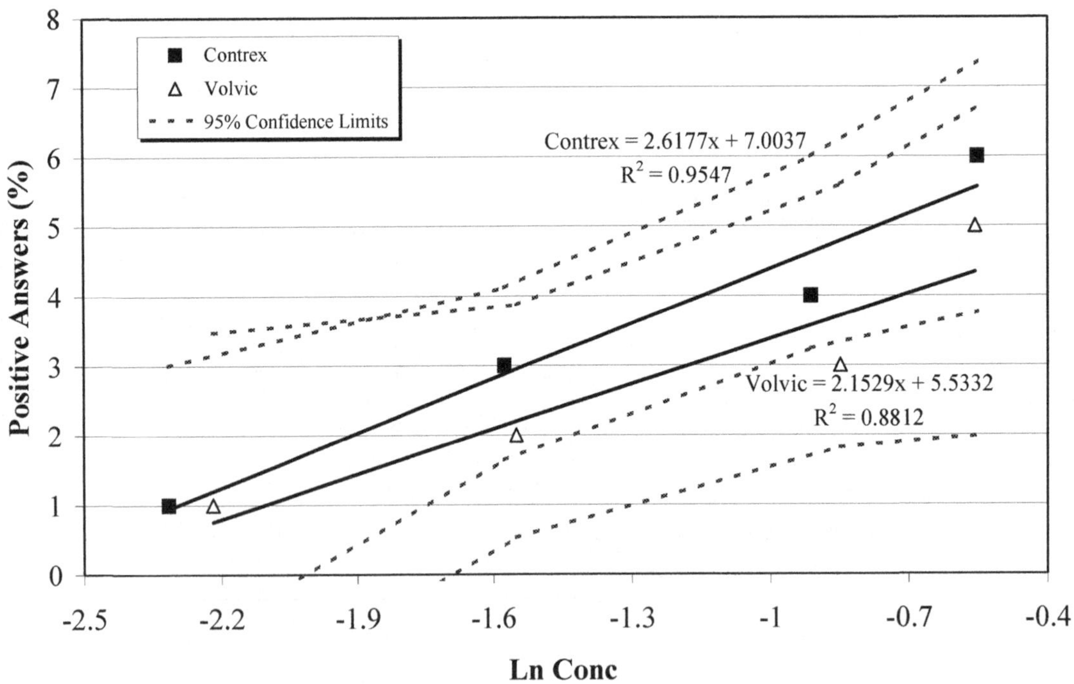

Figure 6.16 Impact of TDS on the median perception of free chlorine by trained panelists

Influence of Earthy-Musty Tastes

The impact of earthy/musty tastes on perception of free chlorine was assessed by the trained panelists. Figure 6.17 illustrates the results obtained using water with a FTN of 3.

The presence of a background musty taste in the chlorine dilution solution induced a shift in the perception of the chlorinous flavor – there was a significant increase in the chlorine threshold sensitivity (0.05 to 0.1 mg/L Cl_2). After dechlorination with thiosulfate, the initial musty-taste solution had the same FTN.

After this first experiment, free chlorine solutions were prepared with different levels of musty taste (FTN= 5, 10, and 20) and submitted to the panelists. Figure 6.18 shows the influence of the increasing intensity of the background musty taste on the perception of free chlorine. Increasing the musty background taste induced an antagonistic shift of the perception of free chlorine, as previously noticed in the seven-market "taste tests." Such antagonistic effect between musty and chlorine taste was also reported by Zhang et al. (1992a, 1992b) and Schweitzer et al. (2002) with 2-methyl isoborneol (MIB). A linear relationship was found between the musty flavor threshold and the free chlorine taste threshold (Fig. 6.19). Schweitzer et al. (2002) also showed both antagonism and synergism could occur at different concentrations ratios of earthy/musty and chlorine compounds.

It is often discussed in the literature that chlorine may have a masking effect on the perception of other tastes present in the water (Figure 6.20), but the contrary can also be true, as showed by these experiments.

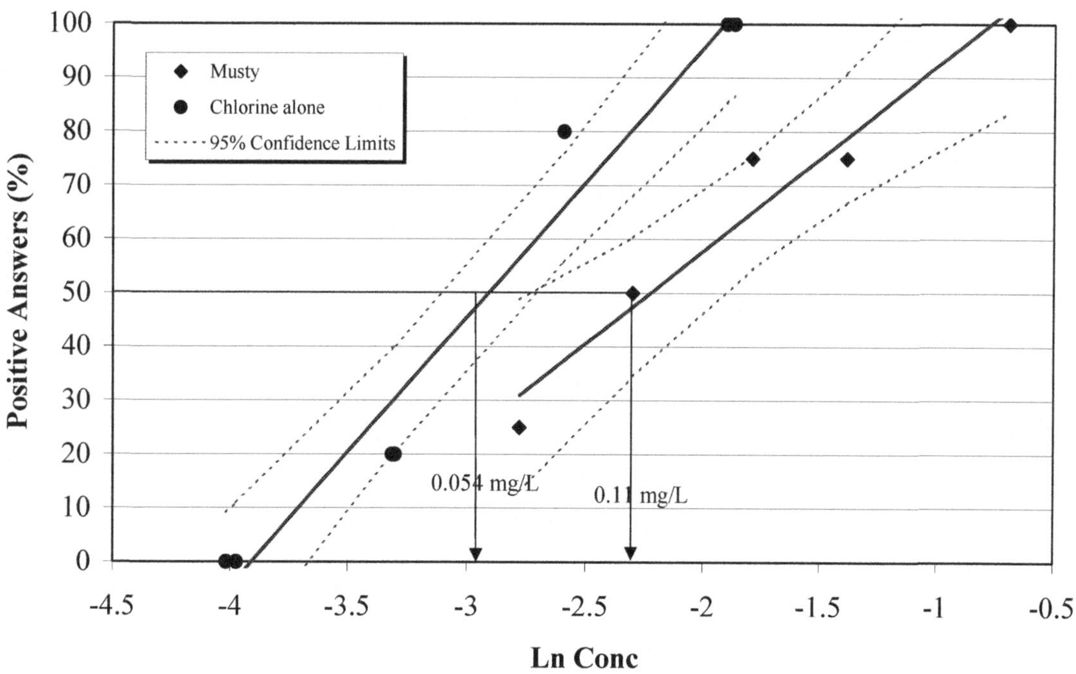

Figure 6.17 Impact of background musty taste in the test water (FTN musty =3) on perception of free chlorine by trained panelists

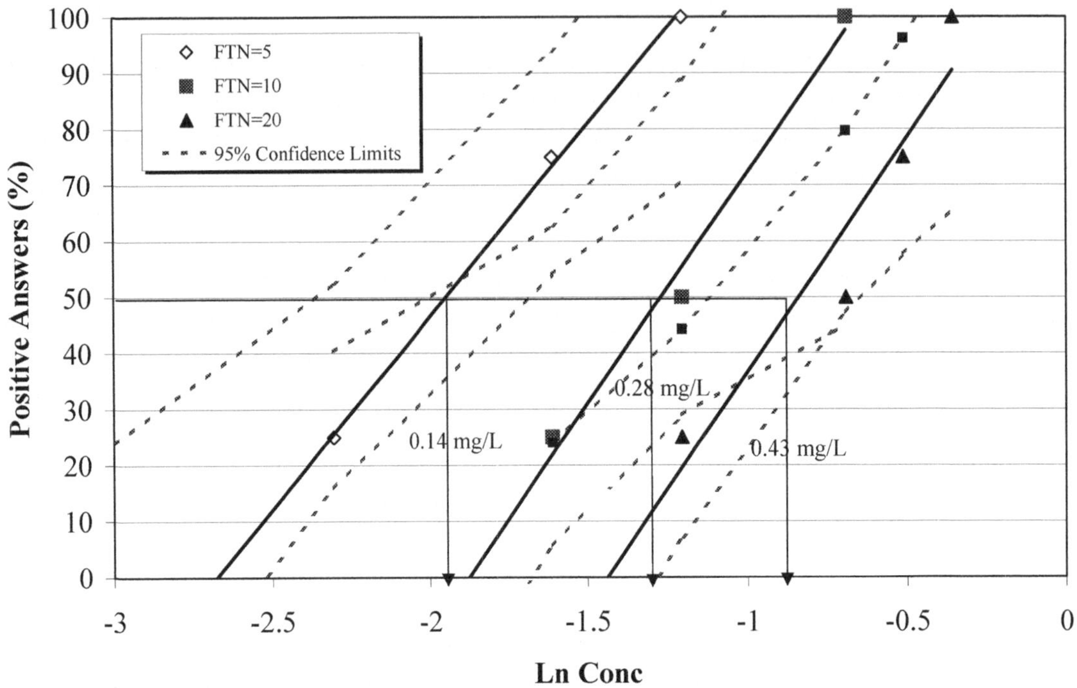

Figure 6.18 Impact of earthy/musty taste on perception of free chlorine by trained panelists

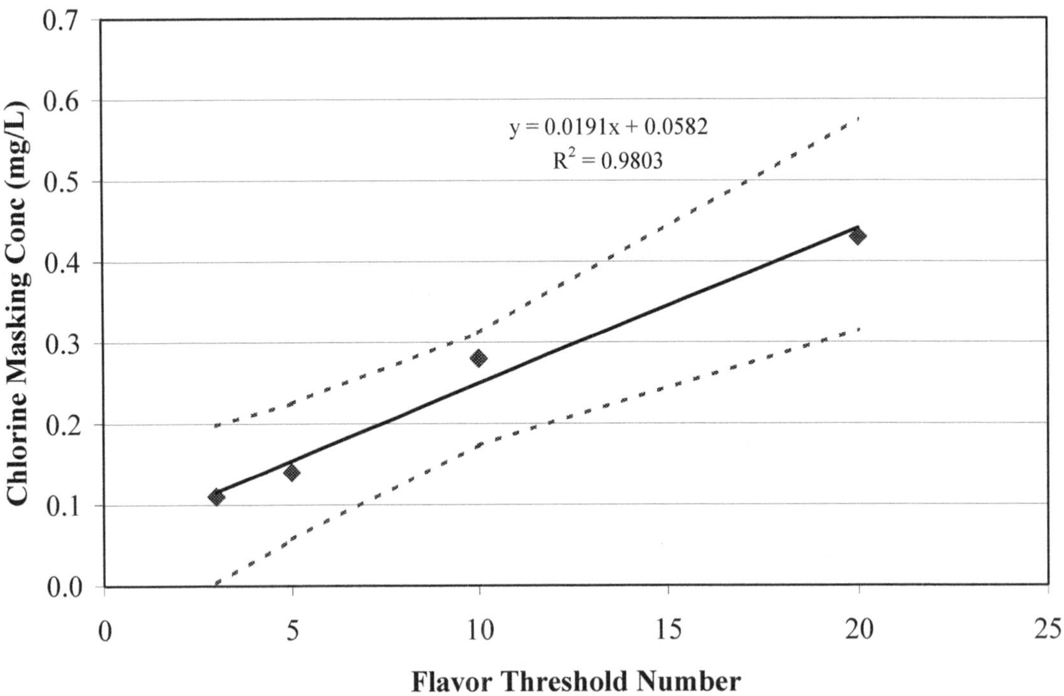

Figure 6.19 Relationship between chlorine perception and the level of background musty taste in the test water

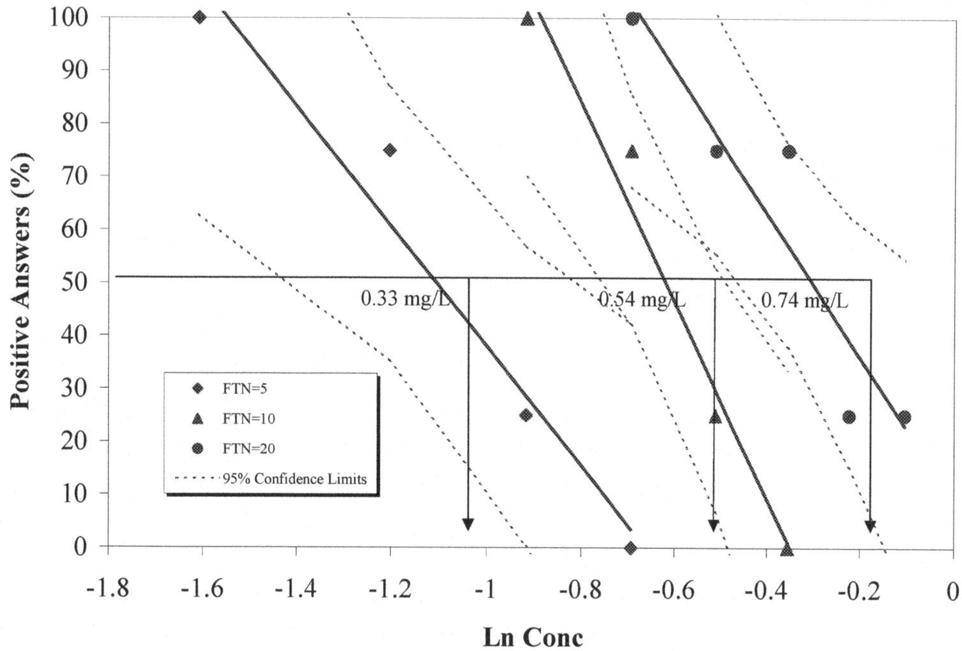

Figure 6.20 Masking effect of chlorine on musty taste perception in the test water (TT musty =5, 10 and 20) by trained panel

Summary of Results

Extensive taste tests were conducted both in France and in the U.S. to determine the detectable and acceptable levels of the main disinfectants in tap water (i.e., chlorine, chloramines, and chlorine dioxide).

Flavor threshold concentrations were first reassessed by trained panelists (4-5 experts). Threshold sensitivities of 0.05 mg/L Cl_2, 0.10 mg/L Cl_2 and 0.2 mg/L ClO_2 were measured for chlorine, monochloramine and chlorine dioxide, respectively (Table 6.13). These thresholds were significantly lower than those reported in the literature (except for chlorine dioxide).

Table 6.13
Flavor threshold of disinfectant by trained panel

Disinfectant	FPA	Triangle test
Chlorine (mg/L Cl_2)	0.06 ± 0.03	0.05 ± 0.02
Monochloramine (mg/L Cl_2)	0.145 ± 0.03	0.085 ± 0.06
Chlorine dioxide (mg/L ClO_2)	0.2 ± 0.07	0.18 ± 0.03

A group of ten average French consumers was 2 and 4 times less sensitive than the trained panelists to monochloramine and free chlorine, respectively (Table 6.14). The answers of untrained panelists were also more heterogeneous than those of trained panelists, as expected.

Each disinfectant's concentration had to be significantly increased to be detected by the majority of the panelists.

Table 6.14
Flavor threshold of disinfectant by untrained panel

Disinfectant	French panel	U.S. panel
Chlorine (mg/L Cl_2)	0.2 ± 0.05	0.9 ± 0.25
Monochloramine (mg/L Cl_2)	0.175 ± 0.05	1.8 ± 0.4

Both untrained panels had similarly heterogeneous responses among their panelists, but the U.S. consumer panel was 5 and 10 times less sensitive than the French consumer panel to free chlorine and monochloramine, respectively. This result is likely linked to the difference in chlorine disinfectant residual concentrations practices in the two countries (0.1 - 0.2 mg/L Cl_2 in France versus ~1.0 – 1.5 mg/L Cl_2 in the U.S.).

Taste testing experiments confirmed that untrained average consumers were able to distinguish between different flavor descriptors, but were not able to describe them correctly.

Potential influences of water characteristics were investigated. No significant influence of temperature (in a range between 15 – 25°C) was noticed on perception of free chlorine. Some influence of TDS on perception of free chlorine was seen, but the results did not provide a clear answer. As described elsewhere, it was confirmed that the presence of musty taste may have a significant negative impact on sensitivity to chlorine.

CHAPTER 7
UTILITY MANAGERS' SURVEY

INTRODUCTION

How well do utility managers know their customers? As previously discussed, consumers reported "chlorinous" flavors in their tap water, although their threshold sensitivity was higher than the residual present, and therefore, objectively impossible to detect them. Do your customers associate "chlorinous" flavors with safe or unhealthy water? Knowing customers is the key to develop public relations (PR) projects that can address their concerns and issues and develop treatment strategies that may achieve better results.

OBJECTIVE

The objective of this chapter is to present an evaluation of a manager's survey that was designed to assess utility managers' knowledge of their customers.

METHODOLOGY

A survey very similar to the consumer survey, but significantly shorter, was sent to over 120 utility representatives, 100 in the U.S. (Figure 7.1) and 23 worldwide (Figure 7.2). The utility managers' survey was distributed to water agencies nationwide, covering 22 states located in various regions throughout the country. Personal contacts and referrals from participating utilities were used to recruit the survey participants. Efforts were made to survey a broad range of managers both geographically and by water treatment plant size and type, but as this was not the main focus of this work, this survey was not meant to be a comprehensive characterization of "utility manager perception in the United States;" instead it was intended to serve as a first look at utility manager attitudes.

Survey questions roughly followed the same format as for the consumer survey: the first set of questions focused on tap water quality, the second set focused on utility service. Utility managers were asked to respond to questions from the viewpoint of their customers. For example, questions like: "What is your overall satisfaction with the quality of tap water in your home?" "What is your average customer's level of satisfaction with the overall quality of tap water in your household?" Surveys were e-mailed or faxed and results hand-entered into Excel™ for further statistical analysis.

The survey was designed to assess utility manager perception of consumers and consumer attitudes toward utilities and to complement the data collected in the customers' survey. While the survey allowed interviewees to respond openly, it was designed to require less than 10 minutes to complete. To promote candid responses, utility managers were informed that the results would be published anonymously. A copy of the survey questions is provided in Appendix A.

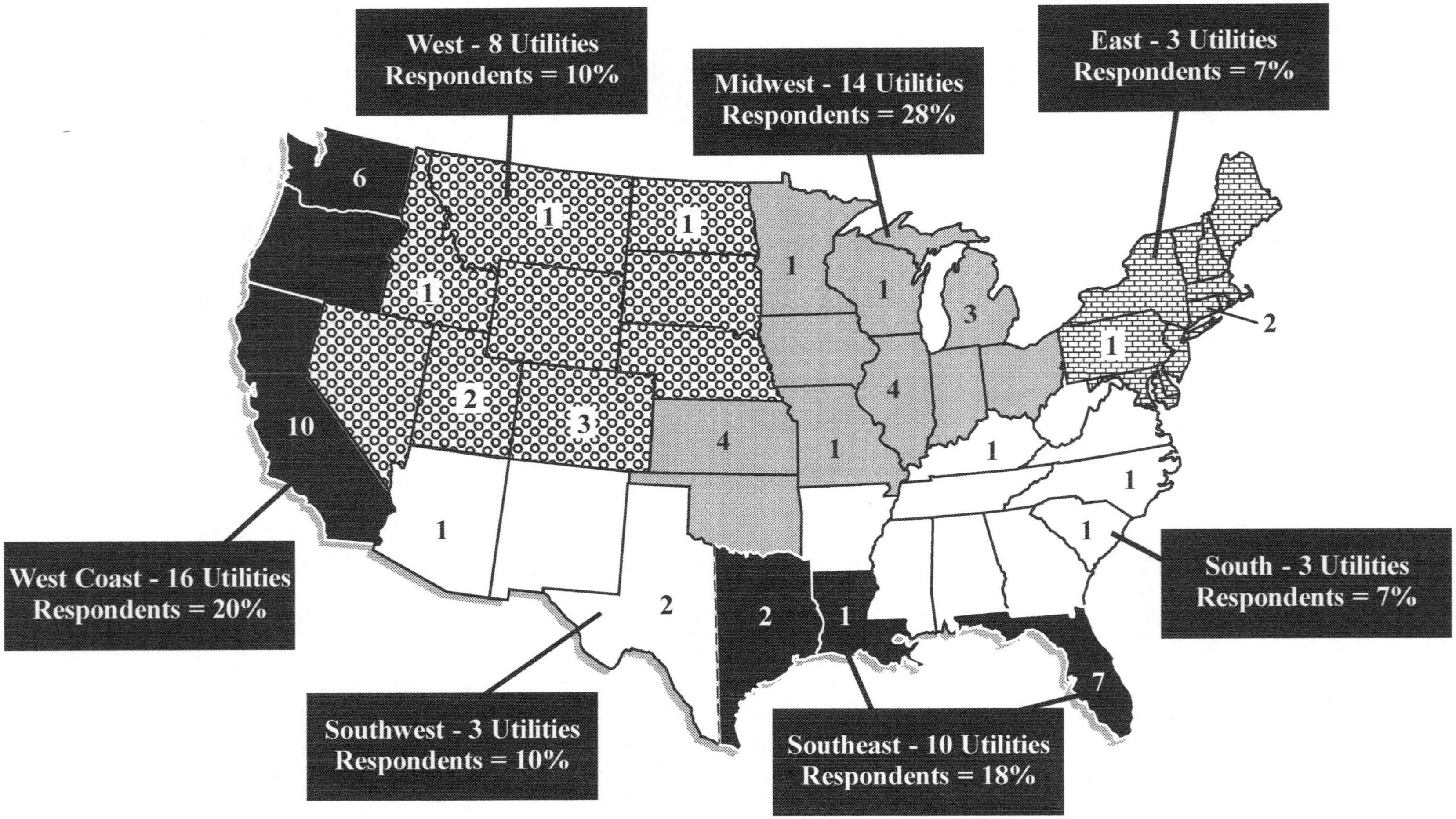

Figure 7.1 Geographic distribution of U.S. utilities participating in the survey

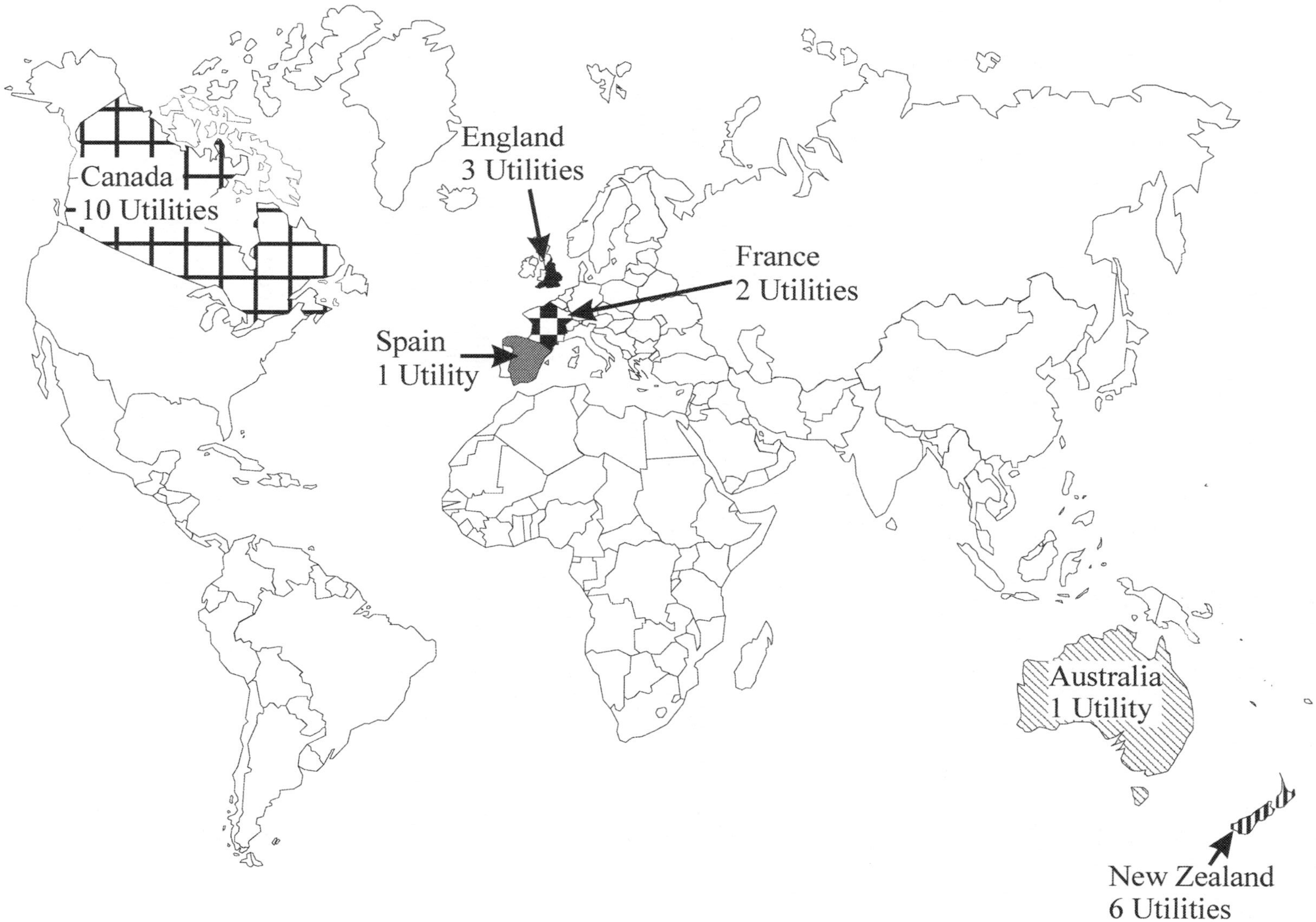

Figure 7.2 Geographic distribution of international utilities participating in the survey

Results showed that utility perceptions of their consumers' perceptions were not always in line with the measured consumer views. These incongruences were found in areas of water quality, utility service, and motivators for use of tap water alternatives. Furthermore, the usefulness of CCRs was overestimated.

UTILITY DEMOGRAPHICS

This chapter first details the demographic background of national and international utilities participating in the survey; after that, utility and consumer responses are compared with specific focus on water quality and utility service aspects; and, finally, the perceived effectiveness of CCRs as a public outreach tool is examined. Throughout discussions, comparisons are made between U.S. utility managers, international utility managers, and U.S. consumers.

A total of 116 responses were received from U.S. agency representatives associated with 57 utilities (Figure 7.1). Surveys were completed by utility directors, superintendents, operators, and service engineers.

Figure 7.3 illustrates the breakdown of surveyed utilities in terms of treatment capacity, raw water source, and respondent job description. Larger-sized utilities (> 20 mgd) were over-represented as compared to smaller ones (86% versus 14%). However, since 81% of the U.S. population is served by only 2% of the existing U.S. utilities (EPA 2001), this uneven distribution was not detrimental to the study. On the contrary, it represented the utilities supplying water to the large majority of consumers, and thus, those who can impact them the most.

Utility diversity based on source water (ground vs. surface water) followed similar trends for community water systems in the U.S. as reported by the EPA (70% and 30% for surface and ground waters, respectively, EPA 2001). In the manager's survey, the representation of surface and ground waters was 60% and 40%, respectively. One measure of diversity, utility ownership, was obviously skewed in favor of public utilities, with only 6% of the responses coming from private utilities.

The intent of the managers' survey was to ascertain the viewpoint of those in positions which could affect consumer attitudes/perceptions. While operators and service engineers could indeed have important roles in the operation of a water utilities, typically those in higher, management level positions (e.g., plant managers or department directors) have greater political leverage to make long-term operational or management changes happen. Overall, nearly two-thirds (64%) of responses came from people in these managerial positions, and therefore, the survey represents the perception of people that could more forcefully impose their views.

Twenty-three utility representatives from five different countries participated in the international portion of the managers' survey (Figure 7.2), four of which were privately owned utilities. As in the U.S. market, Figure 7.4 shows a classification of the international utilities according to plant capacity, raw water source and respondent position. Again, larger utilities (> 20 mgd) were more represented than smaller ones (66% versus 17%). The only significant difference between the U.S. and international background was the number of "did not answer" (DNA) respondents, and is likely an artifact of the sampling method. Ease of correspondence and more than one respondent per utility in the U.S. survey allowed for a more complete demographic database to be developed for each utility.

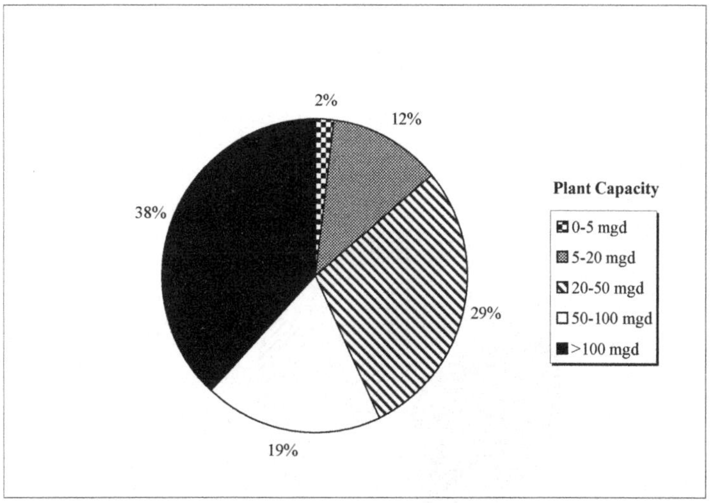

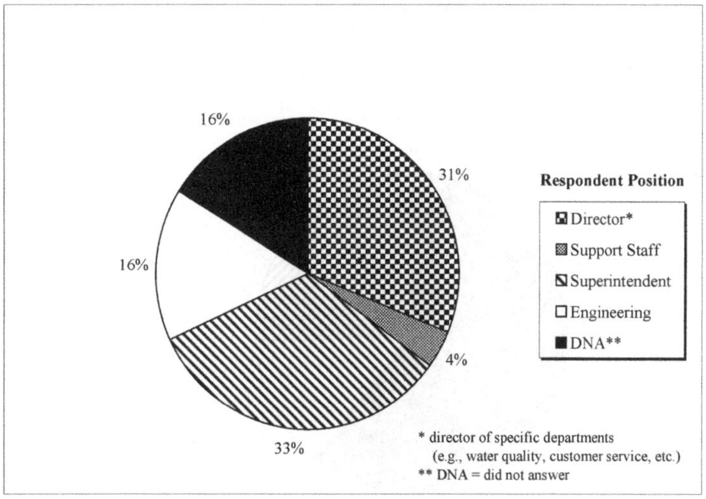

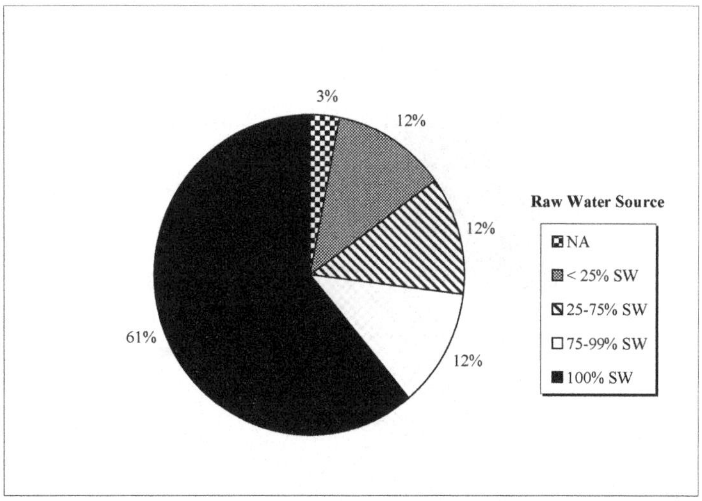

Figure 7.3 Classification of surveyed U.S. utilities by plant capacity, raw water source, and respondent position

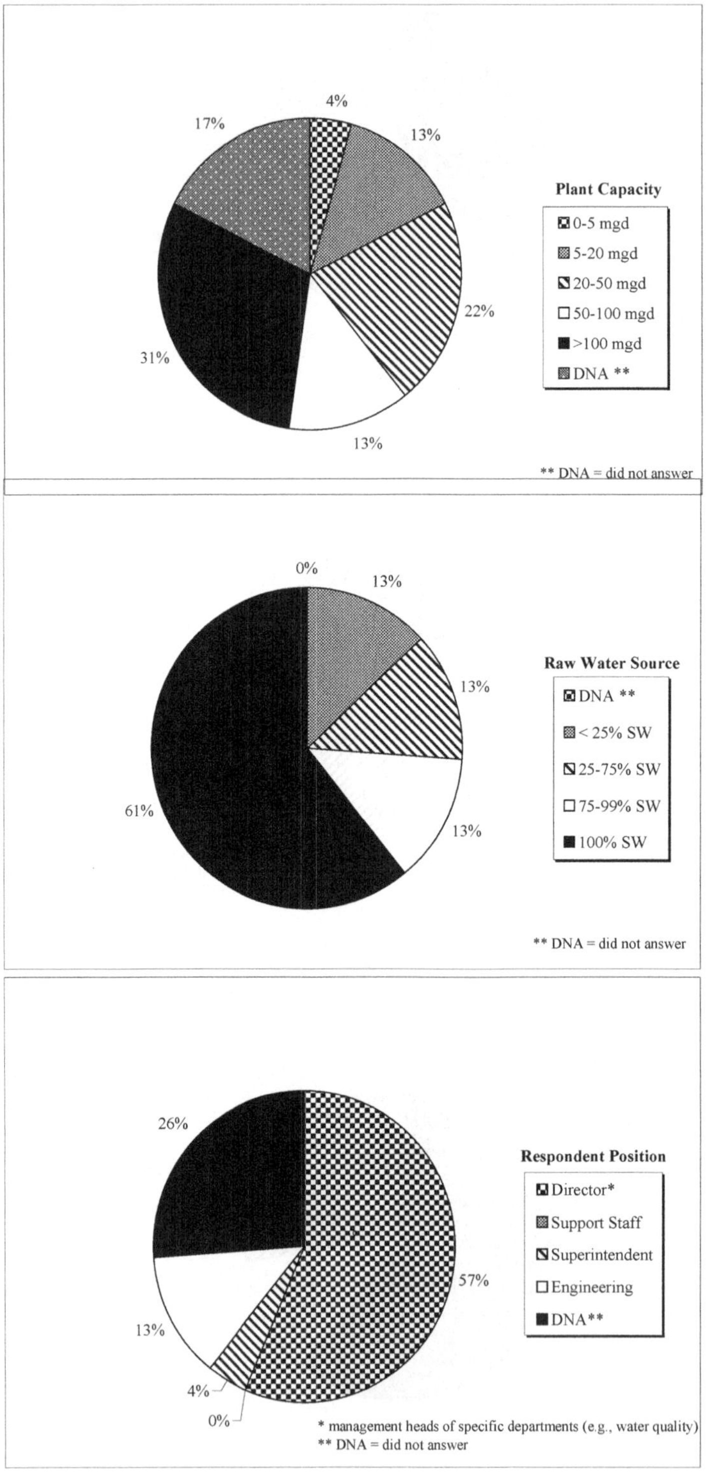

Figure 7.4 Classification of surveyed international utilities by plant capacity, raw water source, and respondent position

MANAGERS SURVEY EVALUATION

Utility managers were asked to rate their average consumers' satisfaction with the overall quality, taste, and safety of their tap water. Table 7.1 summarizes their answers.

Table 7.1
Managers' perceptions of consumer satisfaction with water quality

Level of satisfaction	Percent respondents per water quality feature		
	Overall quality (Q1)	Taste (Q2)	Safety (Q3)
Very satisfied	23	15	61
Satisfied	67	61	28
Neither dissatisfied or satisfied	8	20	5
Dissatisfied	2	4	5
Very dissatisfied	0	0	0
Did not answer	<1	<1	1

Figure 7.5 shows a comparison between consumer and utility manager answers in terms of tap water overall quality, taste, and safety. These results were based on the 392 respondents in the national consumer survey ("Consumer Response") and 57 in the utility survey ("Utility Response"). Categories that do not sum to 100% were due to respondents who refused to answer the question.

Chlorinous Taste and Odor Features

The presence of chlorinous taste and odor in tap water was viewed by most managers to be an insignificant problem. Approximately three-quarters of all U.S. and international managers believed that consumers feel safe when they detect a chlorinous T&O in their tap water (Figure 7.6), although on average, 43% of consumers (ranging from 38 to 53%, depending on the specific market) associated chlorinous flavors with lack of safety (Figure 7.7). Furthermore, U.S. managers tended to overestimate how well the consumers were educated about chlorine in drinking water; over 50% believed consumers appreciate the benefits of chlorine. Interestingly, very few consumers (14%) could articulate the basic processes (e.g., filtration, disinfection) used to treat drinking water.

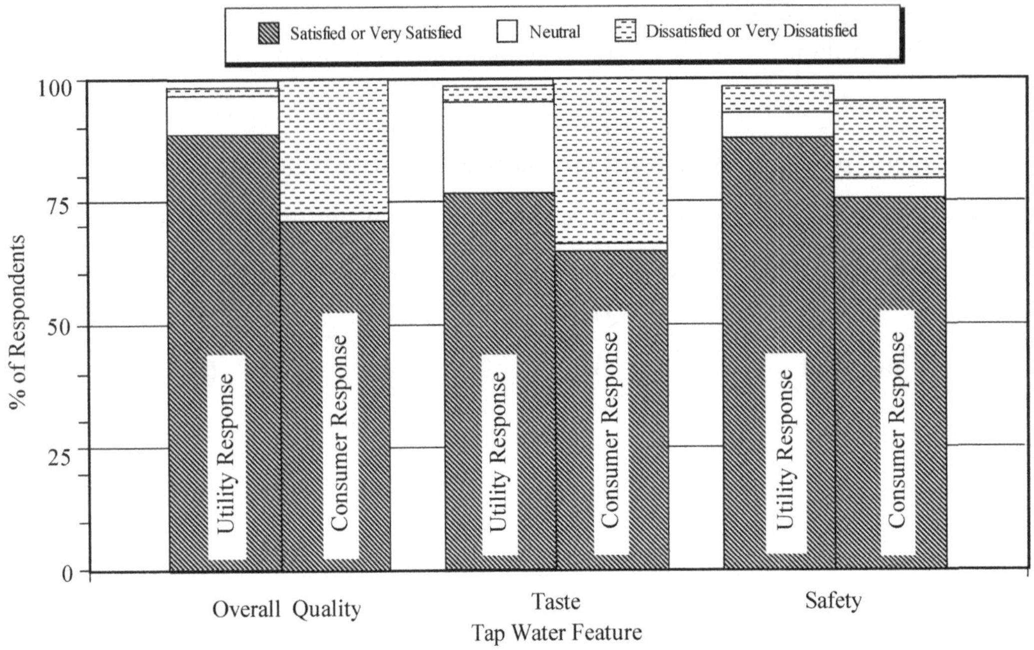

Figure 7.5 Comparison between utility managers and consumers regarding tap water quality features

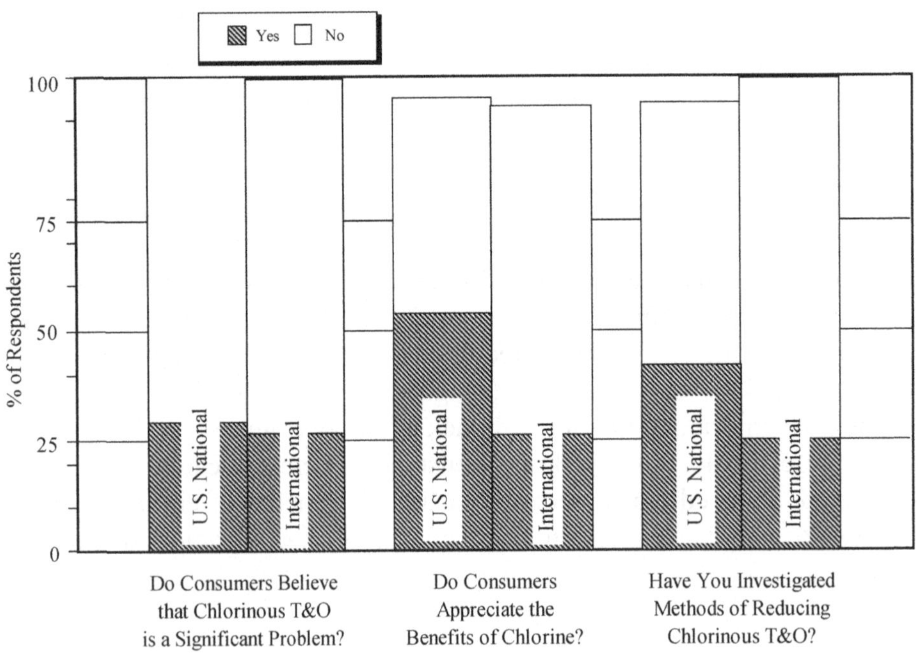

Figure 7.6 U.S. national and international managers' responses to various chlorinous T&O questions

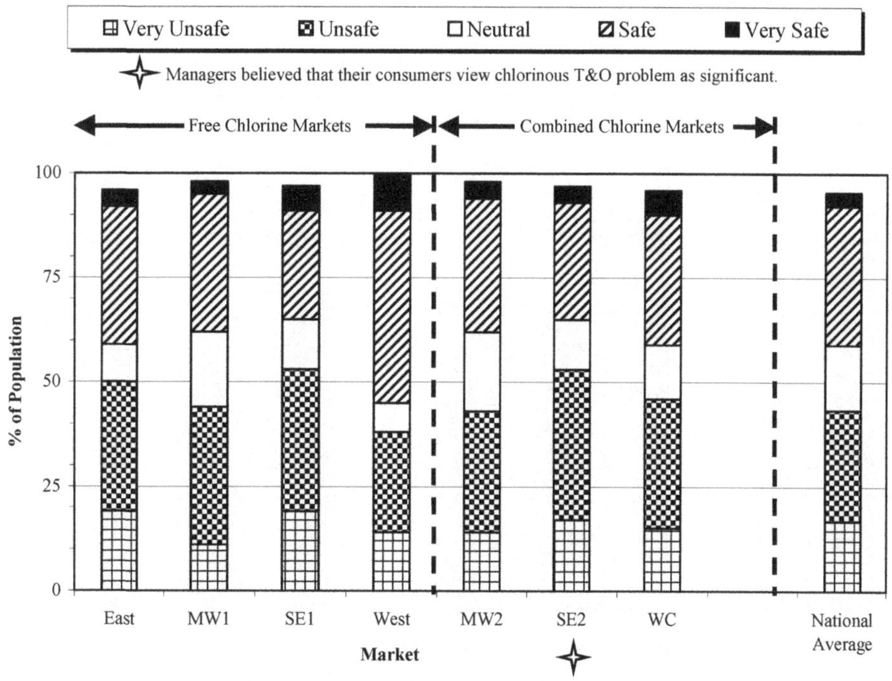

Figure 7.7 Market distribution of consumers citing concerns when chlorine is detected in tap water

Most U.S. managers (53%) believed that consumers understood the benefits of chlorine, but at the same time realized its aesthetic problems need to be mitigated by taste and odor control methods (Figure 7.6). International managers, who are accustomed to either very low chlorine residuals or entirely different disinfection processes (e.g., ozone, ultraviolet radiation), did not express as much concern for chlorinous taste and odor-related issues. This might be due to their limited use of chlorine as a primary disinfectant and greater tendency to use more high tech processes to remove NOM and other potentially odor-causing constituents from the water.

Consumer Satisfaction with Utility Service

Across various levels of utility service, managers consistently perceived their utility as providing exceptional service. Indeed, about three-quarters of the public (in the National Survey) were satisfied with the overall quality of service, but the remaining one-quarter is still a sizeable fraction of the population (Figure 7.8). More revealing was the percentage of people who felt that their complaints or questions had not been properly resolved by the utility. While only a small percentage of the respondents (less than 10%) reported bringing a complaint to their water utility within the last year, only about half of those who did so thought they have received adequate attention. Clearly there is some discrepancy between the service utility managers perceive their water utility to be providing and the perceptions of a significant number of their customers.

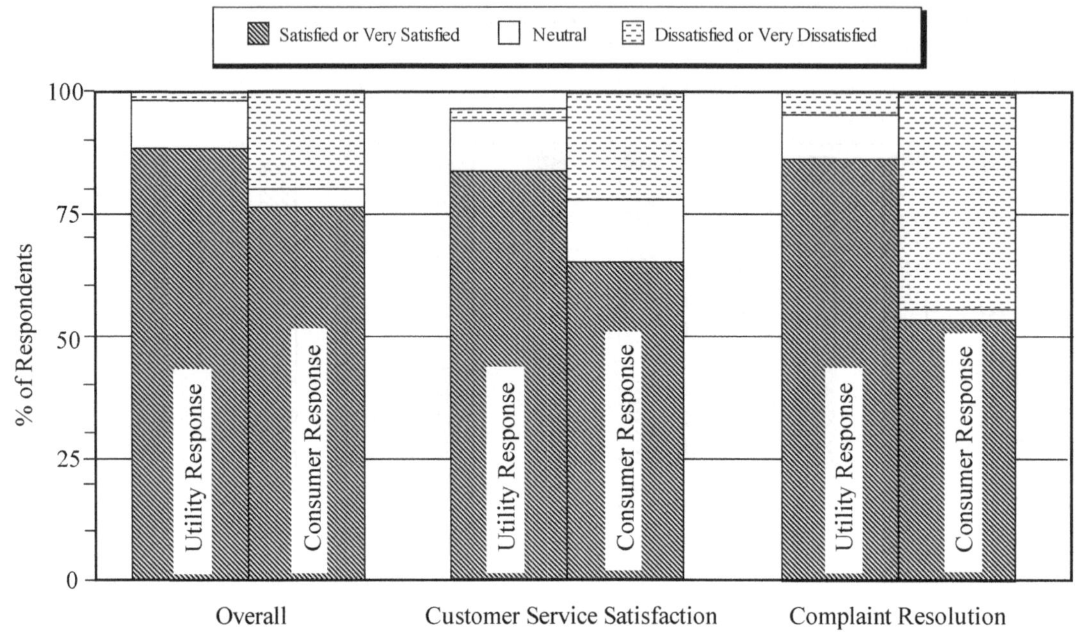

Figure 7.8 Comparison between utility managers and consumers regarding utility service features

Consumer Confidence Reports

Federal regulations require that annual reports regarding tap water quality and safety are to be sent to consumers every year. In order to better ascertain the true effectiveness of CCRs, all consumers were asked the following question "Within the last year, have you received an Annual Drinking Water Quality Report from your local or municipal water provider?" Consumers who did remember receiving a CCR were asked follow-up questions regarding how useful or persuasive was the report. The same questions were included in the utility managers' survey to gauge their evaluation of their consumers.

In spite of the fact that all seven markets participating in the survey were in compliance with the CCR Rule, only 27% of consumers remembered receiving one. Only one utility manager of the seven tested markets believed that the average consumer would recall receiving a CCR (Figure 7.9). Of those consumers that recall receiving a CCR, less than 25% found the report useful. However, in the U.S. 72% of utility managers (national survey) believed that CCRs were a useful method to educate the public.

CCRs were initially intended to better educate the public not necessarily "win them over." Nevertheless, the report can be a very useful marketing tool. However, when consumers were asked if CCRs changed their views about tap water, less than 5% noted that their confidence in tap water safety had increased (Figure 7.10). Evidently, the current format of the report is typically not a very effective marketing tool. A new approach to public outreach needs to be considered. Some new and different outreach methods suggested by utility managers included special newsletters, newspaper ads/articles, materials developed for use in schools, and community events/contests sponsored by the utility.

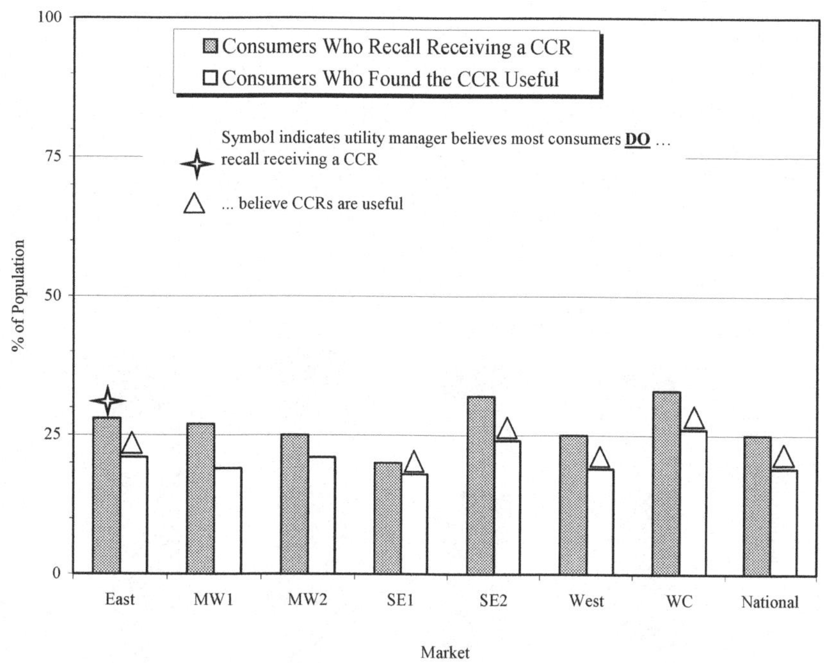

Figure 7.9 Consumers who recall receiving a CCR/Annual Water Quality Report and who also found the report useful

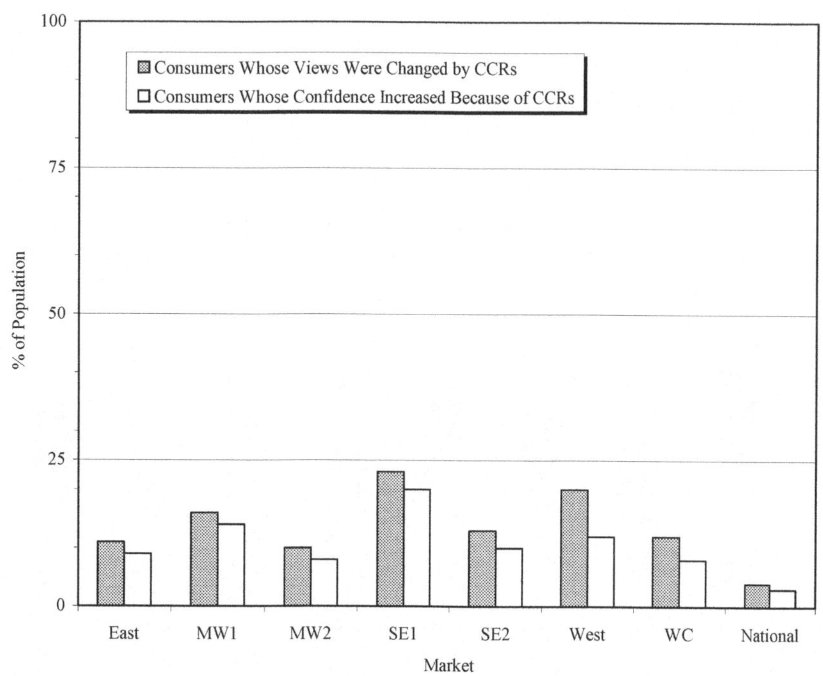

Figure 7.10 Consumers whose views were changed by the CCR and their associated increase or decrease in confidence as a result

Motivators for Bottled and Filtered Water Use

Utility managers were asked to select from six water quality issues the most common reason for customers to choose to use bottled water and filtered water (Figure 7.11). Taste and overall quality were most commonly cited among U.S. and international managers. These two parameters were also the most commonly cited by the consumers in each of the seven markets surveyed (Figure 7.12). However, in their respective markets, only three managers identified taste as the most critical water quality parameter that motivated consumers to switch to bottled water.

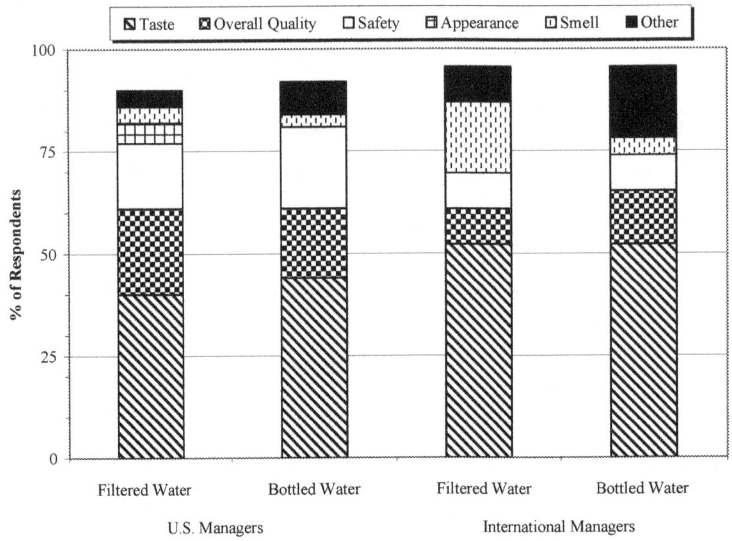

Figure 7.11 Primary motivators for the use of bottled and filtered water as perceived by domestic and international utility managers

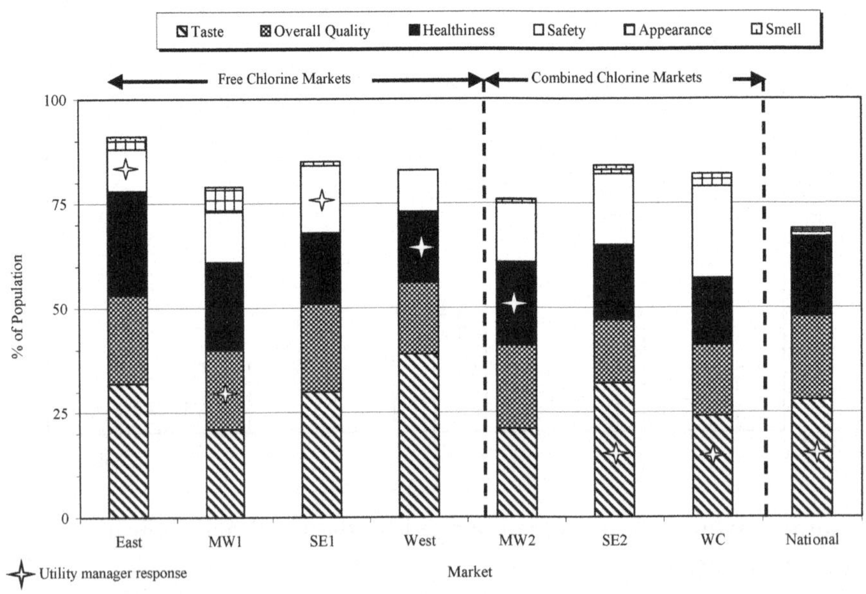

Figure 7.12 Market comparison between the utility and consumers regarding primary motivators for bottled water use

When it came to the use of filtration devices, again utility managers views were not in line with their consumers. Only two identified overall quality as the source of concern that let consumers to opt for a filtration device, though this was, in general, one of the most oft-cited reasons (Figure 7.13).

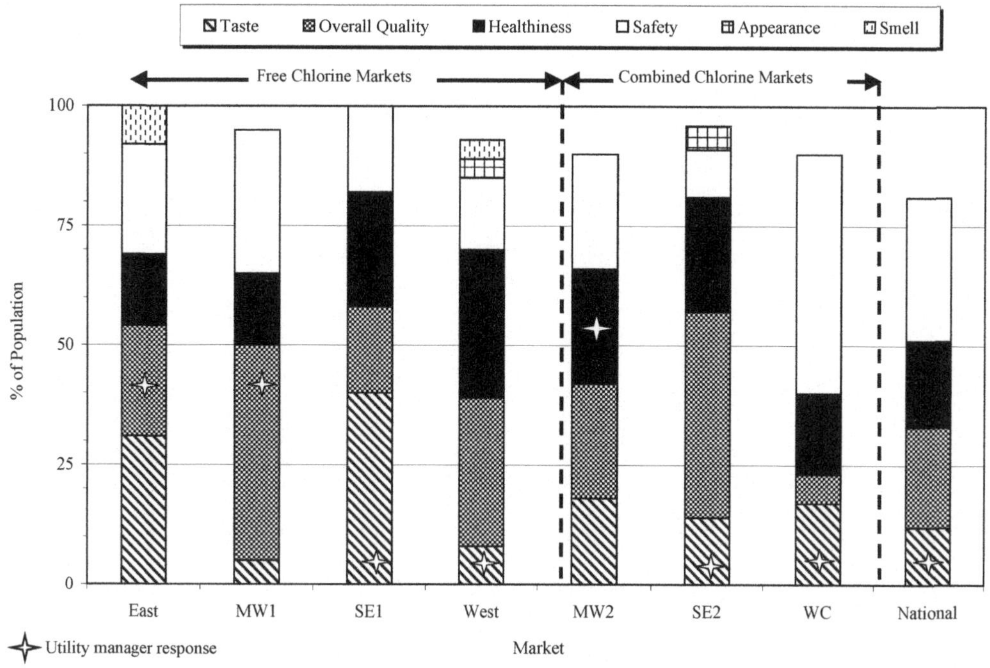

Figure 7.13 Market comparison between the utility and consumers regarding primary motivators for use of a filtration device

CHAPTER 8
FINDINGS AND RECOMMENDATIONS

FINDINGS

The following questions were posed at the beginning of this research project. They are summarized below with the answers developed in this work. Chapter references are provided to guide the reader on where to find detailed discussion of the issue:

1. *What is public perception of tap water chlorinous flavors? Does a significant fraction of the public perceive chlorinous flavor in their water? Does it affect their level of satisfaction with water quality? What is their sensitivity to chlorinous flavor? Are more sensitive individuals less satisfied?* (Chapters 2 - 5)

Results gathered during this study showed that most consumers were satisfied with their tap water, although perception of an off flavor (chlorinous or otherwise) decreased consumer satisfaction. Decreased satisfaction was often associated with a perception of tap water unhealthiness and lack of safety. There were no statistically significant differences in satisfaction between the consumers in the chlorine markets as a group and the consumers in the monochloramine markets as a group, though there were significant differences in satisfaction between individual markets. This was in line with survey results reported elsewhere (Mackey et al. 2003). In general, greater consumer dissatisfaction was observed in the random National Survey (as compared to the seven target markets) for all parameters measured.

Consumer perception of chlorinous flavors was generally less unfavorable than perception of other off flavors, and consumers were generally trusting of the government and their water utility's ability to provide them with safe drinking water. However, there was still a significant portion of the population (~10%) that did not trust the water industry's ability to provide safe drinking water.

According to the results collected from the national and regional market surveys, poor perception of tap water taste was the dominant motivator that led consumers to switch from drinking tap water to drinking TWAs, such as bottled or filtered water. Since "chlorinous taste" was one of the most commonly cited taste descriptors, it was hypothesized that increased perception of chlorinous taste would motivate consumers to switch to TWAs. To examine this hypothesis, the percentage of consumers that described their tap water taste as "chlorinous" was related to the percentage of TWA users. There was no obvious trend in selection of TWAs as a function of perception of chlorinous flavors in tap water.

Consumer sensitivity to chlorinous flavor in their own tap water was tested in the seven telephone-surveyed target markets. The combined average measured consumer sensitivity to free chlorine was 0.82 mg/L Cl_2, approximately three times the value reported in the literature for a trained panel (0.24 - 0.30 mg/L Cl_2, Krasner and Barret 1984). Only 20% of panelists had a threshold sensitivity at or below the literature value. The average measured consumer sensitivity to monochloramine was 3.67 mg/L Cl_2, more than seven times the value reported in the literature for a trained panel (0.28 - 0.48 mg/L Cl_2, Krasner and Barret 1984 and Piriou and Pirelle 1999). In fact, the literature value threshold was very close to the minimum threshold measured in the combined chlorine markets (0.59 mg/L Cl_2).

Human sensitivity to chlorinous flavors was compared for panelists with different flavor profiling skill levels (Flavor Profile Analysis panelists versus average consumers in France and the United States) and from geographical regions with very different chlorine residual concentration disinfection practices (France versus the United States, typical residuals of 0.1 mg/L Cl_2 versus 1 mg/L Cl_2), using a trained FPA panel and untrained panels in France and in the United States. Although some untrained panelists had sensitivities as low as the trained panelists did; on average, they were less sensitive in both countries. Regarding the relative sensitivity of the U.S. and the French untrained panels to chlorine, the difference between the two panels was mainly a shift in the flavor perception trend. The French panel appeared to be more sensitive to lower concentrations, but the trends in increasing positive perception were the same for both untrained panels.

2. *How does chlorine taste affect public perception of health risk associated with drinking water?* (Chapter 4)

Results collected during this study indicated that consumers often did not accurately identify chlorinous tastes and/or odors they perceived in their tap water. It is speculated that consumers may misidentify some objectionable odors as "chlorinous" because that is one drinking water constituent most people know is present in the water. Regardless of the reason, taste test panelists who perceived no/fresh/good tap water flavors were more satisfied with their water quality at all levels, regardless of their threshold sensitivity than those who perceived off flavors (chlorinous or otherwise). Consumers' perception of chlorinous flavors was often associated with lack of safety and unhealthy water.

3. *To what level are other aspects of the water quality impacting chlorinous perception (TDS, hardness, color, other taste and odors descriptors)?* (Chapter 6)

The impact of temperature, TDS and musty-earthy flavors on chlorinous flavor perception was also evaluated during this study. No impact of temperature and TDS concentration on chlorinous flavors perception of both trained and untrained panelists was observed. The impact of earthy/musty tastes on chlorine perception was addressed by the trained panelists. Results showed that the presence of a background musty taste in the chlorine dilution solution induced a shift in the perception of the chlorinous flavor - there was a significant increase of the chlorine threshold sensitivity (from 0.05 to 0.1 mg/L Cl_2). Increasing the background concentration of the musty compound induced an antagonistic shift of the perception of chlorine flavor - in other words, increasing the musty taste masked the chlorinous taste. Such an antagonistic effect between musty and chlorine taste was also reported by Zhang et al. (1992a, 1992b) and Schweitzer et al. (2002) with MIB. A linear relationship was found between the musty flavor threshold and the chlorine taste threshold.

4. *Do CCRs have a comforting or confusing result, exacerbating the consumer's skepticism?* (Chapter 4)

CCRs are helpful, but utilities need to raise consumer's awareness of them if they are to be an effective marketing tool. Consumers were asked a series of questions to evaluate how informed they felt they were with respect to their tap water. Specifically, they were asked whether they received a "water quality report" (Consumer Confidence Report) within the last year, and if so, whether the report was useful. All utilities surveyed were in compliance with CCR regulations, so presumably all respondents were sent one.

Only 30% of the survey respondents remembered receiving a CCR, indicating that the mechanisms currently used to inform the consumers are insufficient and/or not captivating enough to the general public. Eighty percent of the consumers who remembered receiving the CCR found it helpful and were more positive with respect to tap water aesthetics and safety.

5. *How well do we understand consumer perception of tap water quality? Are we on-target, or are we way off-base?* (Chapter 7)

The presence of chlorinous taste and odor in tap water was viewed by most managers surveyed to be an insignificant problem. Approximately 75% of all U.S. and international managers believed that consumers feel safe when they detect a chlorinous T&O in their tap water, although on average, 43% of consumers (ranging from 38 to 53%, depending on the specific market) associated chlorinous flavors with lack of safety. Most U.S. managers (53%) believed that consumers understood the benefits of chlorine, but at the same time realized its aesthetic problems need to be mitigated by taste and odor control methods. International managers, who are accustomed to either very low chlorine residuals or entirely different disinfection processes (e.g., ozone, ultraviolet radiation), did not express as much concern for chlorinous taste and odor-related issues. This might be due to their limited use of chlorine as a primary disinfectant and greater tendency to use more high tech processes to remove natural organic matter and other potentially odor-causing constituents from the water.

Across various levels of utility services, managers consistently perceived their utility as providing exceptional service. Indeed, about three-quarters of the public (in the National Survey) were satisfied with the overall quality of service. More revealing was the percentage of people who felt that their complaints or questions had not been properly resolved by the utility. Only about half of the consumers who reported raising an issue with their local utility within the last year thought they have received adequate attention.

Utility managers were asked about what their consumers perceived as the most important water quality parameter. Taste and overall quality were most commonly cited among U.S. and international managers. These two parameters were also the most commonly cited by the consumers in each of the seven markets surveyed. However, in their respective markets, only three managers identified taste as the most critical water quality parameter that motivated consumers to switch to bottled water. When it came to the use of filtration devices, again, utility managers views were not in line with their consumers. Only two identified overall quality as the source of concern that let consumers to opt for a filtration device, though this was, in general, one of the most oft-cited reasons.

RECOMMENDATIONS

The tap water alternatives market has made significant advances in the last decade, while, concurrently, the nation's public drinking water has been becoming increasingly more pure. With more federal regulations overseeing tap water quality and safety than regulations set forth for the private bottled water sector, there should be no reason for consumers to doubt the safety of public drinking water.

It has been documented here that:

- Consumers generally have a negative opinion of chlorinous tastes and odors in their drinking water.
- Perception of this flavor leads to lower satisfaction with tap water flavor, healthiness and safety.

- Many consumers may not be sensitive to chlorinous flavors and in many cases this flavor may be misidentified.

These facts indicate that consumer satisfaction can be increased by two methods: 1) improving customer confidence in tap water quality and the safety of chlorinated tap water; 2) reducing chlorinous flavors and thereby reducing their detection.

A number of recommendations for improving customer satisfaction either through process changes or management changes are put forth in the following two sections. Some of these recommendations came from utility managers and a review of the literature.

Public Education Campaigns

While water utilities still operate in a monopolistic market, they must now compete for consumer perception and satisfaction with TWAs. Lost revenue from consumers shifting to bottled/filtered water is infinitesimal and is not a threat to utilities. Less than 5% of all water produced by utilities is actually consumed. However, public image and trust is at stake. In an effort to strengthen public confidence, the federal government has mandated the use of annual water quality reports. While results showed that CCRs were useful to about 30% of the public, the remaining people must be reached by other methods. In many ways, it can be argued that there is little difference between tap water, bottled water, and filtered water in the marketplace. They are all complementary products. Advertising and "selling" public drinking water may be a new concept for water utilities, but to remain in good standing and trust, new approaches and tactics need to be taken.

Marketing specialists attribute as much as 73% of a successful sale due to non-verbal methodologies—i.e., advertisements, attractive logo designs, concern for customer satisfaction (Gallagher 2000). Examples of proactive measures for utilities to implement can range from simple participation in community events to official "media teams" established to brief local news media during emergencies or pertinent current events (Mackey et al. 2003). Water utilities provide a service and product to their customers. By realigning their views, utilities will better be equipped to meet their customer's needs and successfully serve them in an effort to keep a strong customer following.

Some simple customer education efforts are listed below:

- Educate the public about source water protection and drinking water treatment. Chlorination provides important public health protection from microbial pathogens. Water utilities are mandated by the Environmental Protection Agency to maintain low levels of chlorine residuals and to minimize the presence of disinfection byproducts in their distribution systems. Chlorination is practiced very safely in the United States, though many of its citizens may not understand all the issues and health data associated with the presence of chlorine in their drinking water.
  - Public education can be accomplished through tours of the water treatment plant (though heightened safety concerns may make this impossible in some locations), public outreach in local schools and activist groups and many other ways.
- Modify CCRs / Annual Water Quality Reports

- Make chlorine and disinfection by-products safety and health issues a focus. Specifically address consumer concerns about the impact of chlorine on their health.
- Use a marketing approach when designing the report, don't just provide the required tables in an unguided mass transfer of data.
- Advertise or publish them in the local paper, on television, billboards or the radio to increase customer awareness.
- Include a Frequently Asked Questions (FAQ) section to address common customer complaints.

• Have a public relations team and response plan in place to deal with negative "events," be it a boil water notice, a negative media report, or a chemical spill in a nearby area-combat negative press with positive press.
  - It is realized that a utility can only provide information; the utility cannot control how the information is presented. Anecdotal evidence suggests that some media outlets may put a negative "spin" on what a utility presents as positive information to get a larger share of the audience or to increase sales, creating a negative image where a positive one was intended. Methods for developing a proactive, positive news campaign are outside the scope of expertise of the authors. Involvement of public relations experts would likely be well advised before engaging in this type of activity.

Minimize the Presence of Chlorinous Taste and Odor

The data developed in this report strongly suggests that average consumers are very insensitive to monochloramine. This implies that switching from free chlorine to monochloramine should solve most "real" chlorinous flavor taste and odor problems. (A "real" problem is one that is based on consumer detection of, and subsequent negative reaction to, chlorinous flavors.) However, the data collected here indicated that there is no statistical difference between free chlorine market consumers and combined chlorine market consumers in terms of satisfaction with household tap water, confidence that their tap water meets or exceeds all drinking water standards, or consumption of tap water alternatives.

Data also suggests that some of the high measured consumer thresholds may have been due to masking effects from musty or metallic flavors. Improvement in taste and odor treatment to address such problems may create a different taste and odor problem for the utility - increased consumer sensitivity to chlorinous flavors. However, comparison of consumer opinion surveys with their measured thresholds indicates that there was no correlation between threshold and perception of chlorinous flavors, so the effect of such changes in treatment is not completely clear.

It is not expected that switching to monochloramine would solve the problem of customer perception of chlorinous tastes and odors, but it would help utilities eliminate one source of consumer taste and odor perception and would enable the utility to more accurately target the true sources of their taste and odor problem. However, it should be cautioned that switching to monochloramine is not a trivial decision and comes with its own set of operational challenges and potential problems. Some guidelines and information to consider in making such a switch are discussed below.

Chloramines are formed by adding ammonia to a water that contains free chlorine. The ammonia combines with the free chlorine to form a series of combined disinfectants collectively referred to as chloramines. Chloramines have a long history of use in the United States and offer many potential benefits, including:

- Eliminating the chlorinous taste and odors produced by free chlorine.
- Reducing the formation of trihalomethanes and haloacetic acids.
- Providing a more persistent disinfectant residual as compared to free chlorine.

As with any treatment alternative, there are potential disadvantages to the use of chloramines. These include:

- Chloramines are a weaker disinfectant as compared to free chlorine.
- Free ammonia is a nutrient that can promote the growth of nitrifying bacteria.
- Chloramines can be toxic to kidney dialysis patients and to fish.

Although chloramines are a weaker disinfectant than free chlorine, historically they have been considered to be more effective than free chlorine in the control of distribution system biofilms. However, recent evidence indicates that any reduction in biofilms may be only temporary in nature. This is in addition to the potential growth of nitrifying bacteria.

Effective use of chloramines requires establishing a program that takes advantage of the benefits of chloramines while minimizing the drawbacks associated with its use. Specifically:

- Selection of the point of ammonia addition downstream of the point of primary disinfection with free chlorine, ozone, chlorine dioxide or ultraviolet light.
- Design of the ammonia feed facilities and controls to minimize the amount of free ammonia leaving the treatment plant and entering the distribution system.
- Maintaining an adequate chloramine residual within the distribution system.
- Reducing conditions within the distribution system that can promote the decay of chloramines and the subsequent liberation of ammonia.
- Implementation of a distribution system Code of Practice to provide positive control and feedback of water quality within the distribution system.
- Development and implementation of a public notification/information program to avoid potential adverse effects on special water uses such as kidney dialysis, tropical fish, etc.

CONCLUSIONS

So what are the appropriate changes for utilities to make to improve customer satisfaction? More sophisticated, centralized treatment? POU/POE devices operated by utilities? Reduce the disinfectant residual (and, thus, reduction of the disinfectant demand of the water)?

This question has several answers, depending on specific characteristics of each market. For example, more affluent areas with a supportive populace will be able to afford improved treatment, as already occurs in certain California areas, where treatment goals for some parameters are even lower than federally mandated. It will also depend on the demographics and social conscience of the target population. Nevertheless, several strategies could be adopted to meet market demands, some of which are listed below:

- *Openness.* Avoid all knowing authoritatively speaking pamphlets or speakers. Rather evoke open-minded communication. Try an 800 number, or an Internet site where people can get responses in an understandable, easy, fast and open way.
- *Problem solving.* Anticipate that the public will become aware of any potential issue and begin to take measures to deal with it before it becomes a public relations problem. As an absolute minimum, have prepared an intelligent, forthright information piece on the emerging issue.
- *Market possible alternatives.* Whenever perceiving a possible consumer problem, market possible alternatives. For example, if some consumers seem to be particularly dissatisfied with chlorinous flavors in their water, it may be advantageous to propose alternatives such as carbon filters, and explain why they would be more susceptible to higher chlorine levels. If the problem is significant, consider providing filters at competitive rates. Other options include selling bottled water, market POU/POE devices, or home test kits that monitor specific parameters of concern (currently available in baby catalogs).
- *Environmental awareness.* Any water utility anticipating moving into the bottled water business should anticipate how to deal with the question of abandoned plastic bottles. Pick up could be integrated into the system. As one delivers fresh cases of water, take away the old bottles. Waste bottles could become an advertising point in promoting deposited bottled water containers.

APPENDIX A
SURVEY INSTRUMENTS

Telephone Survey Instrument

IntroQ

Hello, my name is <Your Name> calling from Clearwater Research. We are conducting a nationwide study of consumer perceptions of local tap water. This is an important research project, and your cooperation is greatly appreciated.

This telephone number was chosen randomly to be included in the survey.

Is this <repeat phone number >?

1. CORRECT NUMBER	**(GO TO PRIVRES)**
2. NO ANSWER	**(DISP 01)**
3. NORMAL BUS	**(DISP 02)**
4. ANSWERING MACHINE	**(LEAVE MESSAGE USE SCRIPT ONE TIME ONLY)**
5. DO NOT WISH TO DIAL THIS NUMBER	**(NULL ATTEMPT)**
6. NUMBER IS NOT THE SAME	**(GOTO WRONGNUM)**

SHOW ANSWER MACHINE SCRIPT:

Hello, my name is <Your Name> calling from Clearwater Research. We are conducting a nationwide study of consumer perceptions of local tap water. This telephone number was chosen randomly to be included in the survey. I would like to speak with an adult in your household and will be calling back within a couple of days. Your participation is important to the success of this survey. Thank you.

1. LEFT MESSAGE
2. DID NOT LEAVE MESSAGE

PrivRes

Is this a private residence?

1. YES	**(GOTO ADULTS)**
2. NO	**(GOTO NONRES)**

NonRes - **Only get this if PrivRes = 2** (Non-Residential)

Thank you very much, but we are only interviewing private residences.

WrongNum - **ONLY GET THIS IF INTROQ = 6 (NUMBER IS NOT THE SAME)**

Thank you very much. I seem to have dialed the wrong number. It's possible that your number may be called at a later time.

Adults **EVERYONE GETS**

Our survey protocol requires that we randomly select one adult who lives in your household to be interviewed. How many members of your household, including yourself, are 18 years of age or older?

__ ENTER THE NUMBER OF ADULTS

Men **ONLY GET IF ADULTS>1**

How many of these adults are men?

0. None
1. One
2. Two
3. Three
4. Four
5. Five
6. Six
7. Seven
8. Eight
9. Nine

Women **ONLY GET IF ADULTS>1 AND MEN<ADULTS**

How many of these adults are women?

0. None
1. One

2. Two
3. Three
4. Four
5. Five
6. Six
7. Seven
8. Eight
9. Nine

WrongTot **ONLY GET IF ADULTS>1 AND MEN+WOMEN<>ADULTS**

Probe: I'm sorry, something is not right.

Number of men - Number of women - + Number of Adults -

1. Correct the number of men **GO TO MEN**
2. Correct the number of women **GO TO WOMEN**
3. Correct the number of adults **GO TO ADULTS**

Selected – **ONLY GET THIS IF MORE THAN ONE ADULT IN HOUSEHOLD**

The person in your household I need to speak with is the ____________.

Are you the _____________?

1. YES **GO TO YOURTHE1**
2. NO **GO TO GETADULT**

One Adult - **ONLY GET THIS QUESTION IF ONLY ONE ADUL IN HOUSEHOLD**

Are you the Adult?

1. YES
2. NO

GetAdult - **ONLY GET THIS IF ONE ADULT = 2 (NO) OR SELECTED = 2 (NO)**

May I speak with him or her?

1. YES, ADULT COMING TO THE PHONE
2. NO, GO TO THE NEXT SCREEN, PRESS CTRL END AND SCHEDULE A CALL-BACK

DO NOT USE CTRL-END ON THIS SCREEN*

Yourthe1 - **ONLY IF ONEADULT = 1 (YES) OR IF SELECTED = 1 (YES) OR GETADULT = 1**

Then you are the person I need to speak with. Your participation is voluntary, but it is very important because you represent many other people in your community. We will hold your responses in the strictest confidence, but you may decline to answer any question you wish. If you have any questions at any point, please let me know.

This survey covers a wide range of topics concerning your local tap water. It will take approximately 12 minutes, but may take longer based on your answers. I'll read as quickly as I can.

1.	PERSON INTERESTED	**(CONTINUE)**
2.	TERMINATE	**(DISP 06)**

PrescrnQ1

Which of the following best describes how you receive your household water?

Would you say your water comes from . . .

1.	A municipal or public water company	**(GOTO PRESCRNQ4)**
2.	A private water company	**(GOTO PRESCRNQ4)**
3.	A private well on your property	**(TERMINATE DISP 13)**
4.	OTHER:_____________	**GOTO PRESCRNQ2)**
7.	DON'T KNOW / I DO NOT UNDERSTAND	**(GOTO PRESCRNQ2)**
9.	REFUSED	**(TERMINATE DISP 12)**

PrescrnQ2 **(GET IF PRESCRNQ1=4 or PRESCRNQ1=7)**

Do you receive a monthly bill for your household water?

1.	YES	**(GOTO PRESCRNQ1, READ OPTIONS 1 & 2 AGAIN)**
2.	NO	**(GOTO PRESCRNQ3)**
7.	DON'T KNOW / I DO NOT UNDERSTAND	**(TERMINATE DISP 12)**
9.	REFUSED	**(TERMINATE DISP 12)**

PrescrnQ3 **(GET ONLY IF PRESCRNQ1=7)**

Do you currently rent your place of residence?

1. YES	**(GOTO PRESCRNQ4)**
2. NO	**(TERMINATE DISP 13)**
7. DON'T KNOW / I DO NOT UNDERSTAND	**(TERMINATE DISP 12)**
9. REFUSED	**(TERMINATE DISP 12)**

PrescrnQ4

Are you currently an employee of a private or municipal water company, bottled water company, or water treatment manufacturer?

1. YES	**(TERMINATE DISP 14)**
2. NO	**(GOTO QUOTAQ)**
7. DON'T KNOW / I DO NOT UNDERSTAND	**(TERMINATE DISP 12)**
9. REFUSED	**(TERMINATE DISP 12)**

QuotaQ

Which of the following best describes the way you most often drink water in your household.

Would you say you usually drink. . .

1. Tap water straight from the faucet (i.e., unfiltered)	**QUOTA CELL A**
2. Tap water that has been filtered or treated within your household	**QUOTA CELL B**
3. Bottled water	**QUOTA CELL B**
7. DON'T KNOW / I DO NOT UNDERSTAND	**TERMINATE DISP 12**
9. REFUSED	**TERMINATE DISP 12**

[NOTE: IF QUOTA CELL IS FULL, THEN READ FOLLOWING SCRIPT: Thank you for your interest in this research project. We already have enough respondents with that particular drinking preference. Thank you for your time**. TEMINATE: DISPOSITION DISP 11**]

The following set of questions asks about your personal satisfaction with the quality of the tap water you receive in your household.

Let me stress again there are no correct answers. Just pick the answer that best reflects your response to the question.

Q005

Which of the following best describes your satisfaction with the overall quality of the tap water you receive in your household?

Would you say you were. . .

1. Very Dissatisfied
2. Dissatisfied
3. Neither Dissatisfied nor Satisfied
4. Satisfied
5. Very Satisfied
7. DON'T KNOW / I DO NOT UNDERSTAND
9. REFUSED

Q010

Which of the following best describes your satisfaction with the taste of the tap water you receive in your household?

Would you say you were. . .

1. Very Dissatisfied
2. Dissatisfied
3. Neither Dissatisfied nor Satisfied
4. Satisfied
5. Very Satisfied
7. DON'T KNOW / I DO NOT UNDERSTAND
9. REFUSED

Q015

Which of the following best describes your satisfaction with the appearance or look of the tap water you receive in your household?

Would you say you were. . .

1. Very Dissatisfied
2. Dissatisfied
3. Neither Dissatisfied nor Satisfied
4. Satisfied
5. Very Satisfied
7. DON'T KNOW / I DO NOT UNDERSTAND
9. REFUSED

Q020

Which of the following best describes your satisfaction with the smell of the tap water you receive in your household?

Would you say you were. . .

1. Very Dissatisfied
2. Dissatisfied
3. Neither Dissatisfied nor Satisfied
4. Satisfied
5. Very Satisfied
7. DON'T KNOW / I DO NOT UNDERSTAND
9. REFUSED

Q025

Which of the following best describes how safe the tap water you receive in your household is to drink?

Would you say your water is . . .

1. Unsafe
2. Somewhat unsafe
3. Neither unsafe nor safe
4. Somewhat safe
5. Safe

7. DON'T KNOW / I DO NOT UNDERSTAND
9. REFUSED

Q026

Which of the following best describes how healthy the tap water you receive in your household is to drink?

Would you say your water is . . .

1. Unhealthy
2. Somewhat unhealthy
3. Neither unhealthy nor healthy
4. Somewhat healthy
5. Healthy
7. DON'T KNOW / I DO NOT UNDERSTAND
9. REFUSED

Q030

Considering all of these water quality features, which one is the most important to you?

Would you say it was your water's. . .

1. Taste
2. Appearance
3. Smell
4. Safety
5. Healthfulness
7. DON'T KNOW / I DO NOT UNDERSTAND
9. REFUSED

(RANDOMIZE LIST FOR PRESENTATION)

Q035

Do you drink the tap water you receive in your household?

1. YES
2. NO — **SKIP TO Q045**
7. DON'T KNOW / I DO NOT UNDERSTAND
9. REFUSED

Q040

Approximately how many glasses of tap water do you personally drink during a full day? (I.E., AT WORK AND HOME)

1. ______
7. DON'T KNOW / I DO NOT UNDERSTAND
9. REFUSED

Q045

How would you describe the taste of the tap water you receive in your household?

10. NONE/NOTHING
11. STALE/FLAT
12. CHLORINOUS (MEDICINAL, SWIMMING POOL)
13. MUSTY/EARTHY
14. SALTY (SALINE)
15. METALLIC
16. PLASTIC
17. FRESH
18. OTHER:______________
77. DON'T KNOW / I DO NOT UNDERSTAND
99. REFUSED

Q046

Do you find the <RESPONSE FROM Q045 HERE> taste objectionable?

1. YES
2. NO
7. DON'T KNOW / I DO NOT UNDERSTAND
9. REFUSED

Q050

How would you describe the smell of the tap water you receive in your household?

1. NONE
2. CHLORINOUS
3. MUSTY/EARTHY

4. PLASTIC
5. FRESH
6. OTHER:______________
7. DON'T KNOW / I DO NOT UNDERSTAND
9. REFUSED

Q051

Would you say that your water smells <INSERT RESPONSE Q050 HERE>. . .

1. Daily
2. At least once a week
3. At least once a month
4. At least once a year
5. Once in the past five years or less
7. DON'T KNOW / I DO NOT UNDERSTAND
9. REFUSED

Q052

Do you find the <RESPONSE FROM Q050 HERE> smell objectionable?

1. YES
2. NO
7. DON'T KNOW / I DO NOT UNDERSTAND
9. REFUSED

Q055

How would you describe the appearance of the tap water you receive in your household?

1. CLEAR
2. COLORED — **GO TO Q060**
3. CLOUDY (TURBID, UNCLEAR)
4. OTHER:______________
7. DON'T KNOW / I DO NOT UNDERSTAND
9. REFUSED

Q060 **GET IF Q055 = 2**

What color?

1. RED
2. ORANGE
3. YELLOW
4. BROWN
5. BLACK
6. WHITE
7. CLOUDY
8. OTHER:______________
77. DON'T KNOW / I DO NOT UNDERSTAND
99. REFUSED

Q061

Would you say that your water looks <INSERT RESPONSE Q055 HERE>. . .

1. Daily
2. At least once a week
3. At least once a month
4. At least once a year
5. Once in the past five years or less
7. DON'T KNOW / I DO NOT UNDERSTAND
9. REFUSED

Q063

Do you use a water-softening unit for the tap water you receive in your household?

1. YES
2. NO **(GO TO Q065)**
7. DON'T KNOW / I DO NOT UNDERSTAND
9. REFUSED

Q064 **GET IF Q063=1**

How long has the water-softening unit been installed in your household?

1. __________
7. DON'T KNOW / I DO NOT UNDERSTAND
9. REFUSED

Q065

Other than a water softener, **(GET IF Q063=1)**

Do you use a water filtration system for the tap water you receive in your household?

1. YES
2. NO — **SKIP TO Q095**
7. DON'T KNOW / I DO NOT UNDERSTAND
9. REFUSED

Q066 **GET IF Q065=1**

Is your water filter directly connected to the water supply line to your house or is it separate?

1. CONNECTED TO WATER LINE **(GO TO Q070)**
   (E.G., UNDER THE SINK, ON THE FAUCET)
2. SEPARATE FROM WATER LINE **(SKIP TO Q095)**
   (E.G., WATER PITCHER, GRAVITY FILTER)
7. DON'T KNOW / I DO NOT UNDERSTAND
9. REFUSED

Q070 **GET IF Q066 = 1**

How many filtration systems do you have installed in your household?

1. ______
7. DON'T KNOW / I DO NOT UNDERSTAND
9. REFUSED

Q071 **GET IF Q066 = 1**

Did you install the filtration system installed in your household?

1. YES
2. NO
7. DON'T KNOW / I DO NOT UNDERSTAND
9. REFUSED

Q075 **GET IF Q066 = 1**

Is your filtration system used to filter all the tap water you receive in your household or just some?

1. ALL
2. SOME
7. DON'T KNOW / I DO NOT UNDERSTAND
9. REFUSED

Q080 **GET IF Q066 = 1**

Where is your filtration system installed?

1. __________
7. DON'T KNOW / I DO NOT UNDERSTAND
9. REFUSED

(IF Q070 > 1, THEN REPEAT Q080 FOR ALL N)

Q085 **GET IF Q066 = 1**

How long has your filtration system been installed in your household?

1. __________
7. DON'T KNOW / I DO NOT UNDERSTAND
9. REFUSED

Q086 **GET IF Q066 = 1**

In the past year, have you cleaned or changed the filter?

1. YES
2. NO
7. DON'T KNOW / I DO NOT UNDERSTAND
9. REFUSED

Q090 **GET IF Q065 = 1**

Which of the following water quality issues is your primary reason for using a water filtration system?

1. Taste
2. Appearance
3. Smell
4. Health
5. Safety
6. Overall quality
7. OTHER:__________
8. DON'T KNOW / I DO NOT UNDERSTAND
9. REFUSED

(RANDOMIZE LIST 1-6 FOR PRESENTATION)

Q095

Do you drink bottled water while at home?

1. YES
2. NO — **SKIP TO Q110**
7. DON'T KNOW / I DO NOT UNDERSTAND
9. REFUSED

Q100 **GET IF Q095 = 1**

How many glasses of bottled water do you personally consume during a full day (I.E., HOME AND WORK)?

1. __________
7. DON'T KNOW / I DO NOT UNDERSTAND
9. REFUSED

Q105 **GET IF Q095 = 1**

What brand of bottled water do you typically purchase?

1. __________

7. DON'T KNOW / I DO NOT UNDERSTAND
9. REFUSED

Q106 **GET IF Q095 = 1**

Which of the following water quality issues is your primary reason for drinking bottled water?

1. Taste
2. Appearance
3. Smell
4. Health
5. Safety
6. Overall quality
7. OTHER:__________
8. DON'T KNOW / I DO NOT UNDERSTAND
9. REFUSED

(RANDOMIZE LIST 1-6 FOR PRESENTATION)

SECOND SECTION *Relation with Local Water Provider*

The following set of questions asks about your opinions regarding your local water provider.

Q110

Which of the following best describes how confident you are in the safety of the tap water you receive in your household?

Would you say you were. . .

1. Not Very Confident
2. Not Confident
3. Neither Not Confident nor Confident
4. Confident
5. Very Confident
7. DON'T KNOW / I DO NOT UNDERSTAND
9. REFUSED

Q115

Which of the following best describes how confident you are that the tap water you receive in your household meets or exceeds state and federal requirements?

Would you say you were. . .

1. Not Very Confident
2. Not Confident
3. Neither Not Confident nor Confident
4. Confident
5. Very Confident
7. DON'T KNOW / I DO NOT UNDERSTAND
9. REFUSED

Q120

Which of the following best describes your overall satisfaction with your local or municipal water provider?

Would you say you were. . .

1. Very Dissatisfied
2. Dissatisfied
3. Neither Dissatisfied nor Satisfied
4. Satisfied
5. Very Satisfied
7. DON'T KNOW / I DO NOT UNDERSTAND
9. REFUSED

Q130

Which of the following best describes your satisfaction with the overall customer service provided by your local or municipal water provider?

Would you say you were. . .

1. Very Dissatisfied
2. Dissatisfied
3. Neither Dissatisfied nor Satisfied
4. Satisfied

5. Very Satisfied
7. DON'T KNOW / I DO NOT UNDERSTAND
9. REFUSED

Q135

Within the past year, have you personally contacted your local or municipal water provider with a question or complaint?

1. YES
2. NO — **SKIP TO Q145**
7. DON'T KNOW / I DO NOT UNDERSTAND
9. REFUSED

Q140 **GET IF Q135=1**

Which of the following best describes your level satisfaction with the result of your question or complaint?

Would you say you were. . .

1. Very Dissatisfied
2. Dissatisfied
3. Neither Dissatisfied nor Satisfied
4. Satisfied
5. Very Satisfied
7. DON'T KNOW / I DO NOT UNDERSTAND
9. REFUSED

Q145

Which of the following best describes your level of satisfaction with water quality information provided by your local or municipal water provider?

Would you say you were. . .

1. Very Dissatisfied
2. Dissatisfied
3. Neither Dissatisfied nor Satisfied
4. Satisfied
5. Very Satisfied
7. DON'T KNOW / I DO NOT UNDERSTAND
9. REFUSED

Q150

Within the last year, have you received a Consumer Confidence Report or an Annual Drinking Water Quality Report from your local or municipal water provider?

1. YES
2. NO — **SKIP TO Q165**
7. DON'T KNOW / I DO NOT UNDERSTAND
9. REFUSED

Q155 **GET IF Q150 = 1**

Did the material presented in the Consumer Confidence Report or Annual Drinking Water Quality Report from your local or municipal water provider change your views about the safety of your drinking water?

1. YES
2. NO
7. DON'T KNOW / I DO NOT UNDERSTAND
9. REFUSED

Q156 **GET IF Q155= 1**

Did your confidence. . .

1. Increase, or
2. Decrease
7. DON'T KNOW / I DO NOT UNDERSTAND
9. REFUSED

Q160 **GET IF Q150 = 1**

Was the information provided in the Consumer Confidence Report from your local or municipal water provider useful to you?

1. YES
2. NO
7. DON'T KNOW / I DO NOT UNDERSTAND
9. REFUSED

Q165

Are you aware of any consumer education programs initiated by your local or municipal water provider?

1. YES
2. NO
7. DON'T KNOW / I DO NOT UNDERSTAND
9. REFUSED

Q170

Do you know the water source from where your local or municipal water provider receives its water?

1. YES
2. NO — **SKIP TO Q180**
7. DON'T KNOW / I DO NOT UNDERSTAND
9. REFUSED

Q175 **GET IF Q170=1**

Is the water source from groundwater, surface water (e.g., lakes, rivers, and streams), or both?

1. GROUNDWATER
2. SURFACE WATER
3. BOTH
7. DON'T KNOW / I DO NOT UNDERSTAND
9. REFUSED

Q180

Do you know what treatment processes are used by your local or municipal water provider to ensure water quality?

1. YES
2. NO — **SKIP TO Q190**
7. DON'T KNOW / I DO NOT UNDERSTAND
9. REFUSED

Q185

What treatment processes are used by your local or municipal water provider to ensure water quality?

1. _________
7. DON'T KNOW / I DO NOT UNDERSTAND
9. REFUSED

Q190

Within the last year, are you aware of any local incident that seriously threatened the quality of water you receive in your household?

1. YES
2. NO — **SKIP TO Q200**
7. DON'T KNOW / I DO NOT UNDERSTAND
9. REFUSED

Q195 **GET IF Q190 = 1**

Can you briefly describe the nature of this local incident that seriously threatened the quality of water you receive in your household?

1. __________
7. DON'T KNOW / I DO NOT UNDERSTAND
9. REFUSED

Q200

Within the last year, are you aware of any national incident that seriously threatened the quality of water in some other community?

1. YES
2. NO
7. DON'T KNOW / I DO NOT UNDERSTAND
9. REFUSED

Q205 **GET IF Q200 = 1**

Where did this national incident that seriously threatened the quality of water occur?

1. __________
7. DON'T KNOW / I DO NOT UNDERSTAND
9. REFUSED

Q210 **GET IF Q200 = 1**

Can you briefly describe the nature of this national incident that seriously threatened the quality of water in some other community?

1. __________
7. DON'T KNOW / I DO NOT UNDERSTAND
9. REFUSED

Q211

Please indicate your level of agreement with the following statements:

In general, I believe it is safe to drink tap water.

Would you say you . . .

1. Strongly disagree
2. Disagree
3. Neither disagree nor agree
4. Agree
5. Strongly agree
7. DON'T KNOW / I DO NOT UNDERSTAND
9. REFUSED

Q212

I believe it safe to drink water that has a chlorinous taste or odor.

Would you say you . . .

1. Strongly disagree
2. Disagree

3. Neither disagree nor agree
4. Agree
5. Strongly agree
7. DON'T KNOW / I DO NOT UNDERSTAND
9. REFUSED

Q213

Scientists and medical experts have a good understanding of any health risks from drinking tap water.

Would you say you . . .

1. Strongly disagree
2. Disagree
3. Neither disagree nor agree
4. Agree
5. Strongly agree
7. DON'T KNOW / I DO NOT UNDERSTAND
9. REFUSED

Q214 **SKIP IF Q025 = 1 or 2**

My water utility can be trusted to provide water that is safe and healthy to drink.

Would you say you . . .

1. Strongly disagree
2. Disagree
3. Neither disagree nor agree
4. Agree
5. Strongly agree
7. DON'T KNOW / I DO NOT UNDERSTAND
9. REFUSED

Q215

Government, not the individual consumer, is responsible for making sure that tap water is safe to drink.

Would you say you . . .

1. Strongly disagree

2. Disagree
3. Neither disagree nor agree
4. Agree
5. Strongly agree
7. DON'T KNOW / I DO NOT UNDERSTAND
9. REFUSED

THIRD Section *Demographics*

Now, I'm going to ask a set of general questions about your personal characteristics. I understand that occasionally respondents consider some questions sensitive. When I read each one, please answer to the best of your ability.

Q225

Which of the following ranges best describes your age?

[READ CHOICES]

1. Under 30
2. From 30 to 44
3. From 45 to 64
4. 65 or older

7. DON'T KNOW / NOT SURE
9. REFUSED

Q230

[INTERVIEWER: RECORD GENDER, VERIFY IF NECESSARY]

1. MALE
2. FEMALE
7. DON'T KNOW / NOT SURE
9. REFUSED

Q235

Which racial or ethnic background best describes you?

Would you say: White, Black, Asian, Pacific Islander, American Indian, Alaska Native, or Other?

[READ CHOICES]

1. White or Caucasian
2. Black or African-American
3. Hispanic or Latin American
4. Asian or Pacific Islander
5. Native American or Aleut (Al-Oot)?
6. OTHER (Specify?)
7. DON'T KNOW / NOT SURE
9. REFUSED

Q245

How long have you lived in the apartment or household in which you currently reside?

RECORD: XXX

777 = DON'T KNOW / NOT SURE
999 = REFUSED

Q250

What is the highest grade or year of school you completed?

[READ ONLY IF NECESSARY]

1. NEVER ATTENDED SCHOOL OR ONLY ATTENDED KINDERGARTEN
2. GRADES 1 THROUGH 8 (ELEMENTARY)
3. GRADES 9 THROUGH 11 (SOME HIGH SCHOOL)
4. GRADE 12 OR GED (HIGH SCHOOL GRADUATE)
5. COLLEGE 1 YEAR TO 3 YEARS (SOME COLLEGE OR TECHNICAL SCHOOL)
6. COMPLETION OF A 2-YEAR TECHNICAL OR ASSOCIATE DEGREE
7. COLLEGE 4 YEARS OR MORE (COLLEGE GRADUATE)
8. GRADUATE SCHOOL
9. REFUSED

Q255

How many children aged 17 and below reside in your household:

XX [ENTER AMOUNT]

20 = 20 or more
77 = DON'T KNOW / NOT SURE
99 = REFUSED

Q260 **IF Q255=1**

What is the child's age?

XX [ENTER AMOUNT]

77 = DON'T KNOW / NOT SURE
99 = REFUSED

Q265 **IF Q255>1**

What are the children's ages?

XX [ENTER AMOUNTS, SEPARATED BY COMMAS]

77 = DON'T KNOW / NOT SURE
99 = REFUSED

Q266

Have smoked or used tobacco products in the past 30 days?

1. YES
2. NO
7. DON'T KNOW / I DO NOT UNDERSTAND
9. REFUSED

Q305

Including all possible sources of income, which of the following ranges best describes the total income for your household last year?

10. Under $15,000
11. At least $15,000 but no more than $25,000
12. At least $25,000 but no more than $35,000

13. At least $35,000 but no more than $45,000
14. At least $45,000 but no more than $55,000
15. At least $55,000 but no more than $75,000
16. At least $75,000 but no more than $100,000
17. At least $100,000 but no more than $250,000
18. More than $250,000
77. DON'T KNOW / I DO NOT UNDERSTAND
99. REFUSED

CLOSING

Thank you very much for helping us in our study. This research project would not be possible without the help of people like you, and we are deeply grateful for your participation.

MANAGER'S SURVEY INSTRUMENT

Respondent Name and Position:

Utility Name:

Utility capacity:

Source water:

Source water classification:

☐ groundwater ☐ surface water

☐ blended (% of each)

Utility classification:

☐ public ☐ private

<u>Geographical location:</u>

Population served:

Survey Questionnaire:

The following list of questions is intended to identify your perceptions as a utility representative of how your customers feels about their tap water and how they feel about bottled water and filtration devices. These questions have been structured to compliment the data to be collected in the customer telephone survey. The questions have been designed to elicit quick responses. However, please feel free to elaborate on any of the issues, as you feel appropriate.

1. Which of the following do you believe best describes your average customers satisfaction with the **overall quality** of the tap water they receive in their household. Would you say they were…?

   ☐ very dissatisfied
   ☐ dissatisfied
   ☐ neither dissatisfied nor satisfied
   ☐ satisfied
   ☐ very satisfied

2. Which of the following do you believe best describes your average customers satisfaction with the **taste** of the tap water they receive in their household? Would you say that they were…?

   ☐ very dissatisfied
   ☐ dissatisfied
   ☐ neither dissatisfied nor satisfied
   ☐ satisfied
   ☐ very satisfied

3. Which of the following do you believe best describes how **safe** the average customer believes the tap water they receive in their household is to drink? Would you say they believe their water is…?

- ☐ unsafe
- ☐ somewhat unsafe
- ☐ neither unsafe nor safe
- ☐ somewhat safe
- ☐ safe

4. Which of the following do you believe best describes how your customers feel about the **healthiness** of the tap water they receive in their household? Would you say that they believe the water is…?

- ☐ healthy
- ☐ somewhat healthy
- ☐ neither healthy nor unhealthy
- ☐ somewhat unhealthy
- ☐ unhealthy

5. Considering the following water quality features which one do you believe is the most important to your customers? Would you say it was the waters…?

- ☐ Taste
- ☐ Appearance
- ☐ Smell
- ☐ Safety
- ☐ Healthfulness
- ☐ Other ____________________

6. Do you believe there is a significant difference in how people who drink tap water and people who drink bottled water perceive the taste of their tap water?

What about the smell of their tap water?

7. Do you believe there is a significant difference between how tap water drinkers and bottled water or filtered water drinkers perceive the healthiness of the tap water that they receive in their household?

What about the safety?

8. How sensitive are your average consumers to chlorinous taste and odor in their tap water…?

- ☐ sensitive
- ☐ somewhat sensitive
- ☐ neither sensitive nor insensitive
- ☐ somewhat insensitive
- ☐ insensitive

9. Do your consumers believe that chlorinous taste and odor is a significant problem for their tap water?
10. Do other water quality aspects (*e.g.* hardness, color) influence people's perceptions of chlorinous flavors in their tap water?

11. Do you believe your consumers appreciate/understand the benefits of chlorine in tap water, or is public education required/needed?

12. Have you investigated methods of reducing chlorinous taste and odor in your finished water?

    If so, were they feasible solutions?

13. Do you have plans to change/improve public perceptions of the chlorinous taste and odor in their water?

14. Of the following water quality issues, which do you believe is the most common reason for customers to choose to use a water filtration system?

- ☐ Taste
- ☐ Appearance
- ☐ Smell
- ☐ Safety
- ☐ Overall quality
- ☐ Other ______________________

15. Of the following water quality issues, which do you believe is the most common reason for customers to choose to use bottled water?

- ☐ Taste
- ☐ Appearance
- ☐ Smell
- ☐ Safety
- ☐ Overall quality
- ☐ Other ______________________

16. Indicate the categories in which you believe that the bottled water consumers in your service area believe bottled water is better than tap water.

- ☐ Taste
- ☐ Smell
- ☐ Safety
- ☐ Healthiness
- ☐ Quality
- ☐ Cost

17. Which of the following best describes your average customer's overall satisfaction with their municipal water provider? Would you say they were…?

- ☐ very dissatisfied
- ☐ dissatisfied
- ☐ neither dissatisfied nor satisfied
- ☐ satisfied
- ☐ very satisfied

18. Which of the following best describes your average customer's satisfaction with the customer service provided by the water utility? Would you say they were…?

- ☐ very dissatisfied
- ☐ dissatisfied
- ☐ neither dissatisfied nor satisfied
- ☐ satisfied
- ☐ very satisfied

19. Which of the following best describes an average customer's level of satisfaction with the resolution of questions or complaints which they may have proposed. Would you say they were…?

- ☐ very dissatisfied
- ☐ dissatisfied
- ☐ neither dissatisfied nor satisfied
- ☐ satisfied
- ☐ very satisfied

20. Do you believe your average customer is aware that they have received a Drinking Water Quality Report from the water utility within the last year?

Is so, do you believe that they found this report useful?

21. Is the report shipped separately or along with a water bill (circle the appropriate answer)?

Separate Along with bill

22. Do you believe that sending out water quality reports is a useful way to educate the public about their drinking water quality?

23. Do you believe there may be a better way to achieve this aim?

If so, what is it?

24. Within the last year, were there any local incidents that seriously threatened the water quality of the water that customers received in their household?

If so, can you described the nature of this incident?

25. Are you aware of any local incidents within the last year that may have caused customers to perceive a threat to the quality of the water they receive in their household, although there was no actual real threat, as in a negative media report or water quality problems in other places?

If so, can you describe this/these event(s)?

26. Are most customer water quality complaints what you would consider legitimate (*e.g.* not due to problems within the home, misunderstandings about water quality, or the result of perceived problems)?

27. Are you concerned about the public's perceptions of tap water?

What aspects of these perceptions concern you most?

28. Does your utility have other PR projects besides CCRs to actively improve the public perception of your tap water or your utility in general?

29. How would you feel about operating / supplying bottled water or POU/POE devices through your utility?

30. Is this something you would seriously consider initiating if it were shown to be legally and logistically feasible?

Thank you for participating in our survey.

As we stated previously, your responses will be kept totally anonymous and only descriptors such as utility size and water source will be used to describe you. If you have any further questions, or are interested in the outcome of the survey, please feel free to give me a call. Thank you again for your time.

APPENDIX B
SURVEY FINAL DISPOSITION CODES AND RESPONSE RATE CALCULATIONS AND RESULTS

Final Disposition Codes

Final Disposition	Description	Coding Algorithm	Step
23	Disconnect/Non Working	If last disposition in the call history was 05 (Disconnect/Non Working).	**3**
24	No Answer	If last disposition in the call history was 01 (No Answer), 03 (Answering machine), or 04 (Left Message).	**9**
25	Not a Private Residence	If last disposition code in the call history was 09 (Business/Fax) or 19 (Pager).	**4**
26	No Eligible Respondent at Number	If last disposition code in the call history was 13 (Private Well), 14 (Employee of Water Company), or 15 (Doesn't Rent).	**11**
27	No Eligible Respondent During Interview Period	If one disposition code in the call history was 10 (Callback).	**8**
28	Language Barrier	If one disposition code in the call history was 07 (Language / Hearing Barrier).	**7**
29	Interview Terminated in Questionnaire	If one disposition code in the call history was 08 (Midterm Refusal in Questionnaire).	**6**
30	Final Line Busy	If last disposition code in the call history was 02 (Busy).	**10**
32	Refusal	If any disposition code in the call history was 06 (Refusal) or 12 (Refused Screening Question).	**5**
36	Complete	If last disposition code in the call history was 20 (Complete).	**1**
37	Quota Cell Full	If last disposition code in the call history was 11 (Quota Cell Full).	**2**

Note: The assignment of unique Final Disposition codes is based on the stepwise application each of the previously listed algorithms. The specific order for the replacement of last disposition codes with final disposition codes is provided in the step column. Once a final disposition has been assigned, no further changes are made based on subsequent steps.

Response Rate Calculation Formulas

1. Council of American Survey Research Organizations (CASRO) Response Rate Calculation Formula

$$\frac{36}{(27+29+36+37)+\left[(22+28)\times\frac{(27+29+36+37)}{(27+29+36+37)+(26)}\right]+\left[\left[(24+30)\times\frac{(22+27+28+29+36+37)}{(22+27+28+29+36+37)+(23+25)}\right]\times\frac{(27+29+36+37))}{(27+29+36+37)+(26)}\right]}$$

2. Upper Bound

$$\frac{36}{(27+29+36+37)}$$

3. Refusal Rate

$$\frac{32}{(22+23+24+25+26+27+28+29+30+36+37)}$$

Note: The previous formulas reference the final disposition codes presented previously.

Response Rate by Water Utility Market Area

1. East Coast

Household	
No Eligible Respondent at Number	219
Final Termination in Questionnaire	34
Refusal (Eligibility Unknown)	933
Complete	192
Quota Cell Full	19
Total Household	1397
Non-Household	
Not a Private Residence	197
Disconnect/ Non-working	743
Total Non-Household	940
Unknown	
Final No Answer	678
Final Line Busy	51
Language Barrier	56
Don't Know	1
Total Unknown	786
Total Dispositions	3123
CASRO	10.28%
Upper Bound	16.57%
Refusal Rate	30.96%

2. Midwest 1

Household	
No Eligible Respondent at Number	451
Final Termination in Questionnaire	16
Refusal (Eligibility Unknown)	918
Complete	193
Quota Cell Full	6
Total Household	1584
Non-Household	
Not a Private Residence	480
Disconnect/ Non-working	1325
Total Non-Household	1805
Unknown	
Language Barrier	91
Final No Answer	929
Final Line Busy	62
Total Unknown	1082
Total Dispositions	4471
CASRO	9.24%
Upper Bound	17.13%
Refusal Rate	20.89%

3. Midwest 2

Household	
Complete	201
No Eligible Respondent at Number	4
Final Termination in Questionnaire	25
Refusal (Eligibility Unknown)	1146
Quota Cell Full	73
Total Household	1449
Non-Household	
Not a Private Residence	49
Disconnect/ Non-working	557
Total Non-Household	606
Unknown	
Final No Answer	1780
Final Line Busy	61
Language Barrier	40
Don't Know	146
Total Unknown	2027
Total Dispositions	4082
CASRO	6.98%
Upper Bound	14.65%
Refusal Rate	28.69%

4. Southwest 1

Household	
No Eligible Respondent at Number	376
Quota Cell Full	103
Complete	176
Final Termination in Questionnaire	28
Refusal (Eligibility Unknown)	1472
Total Household	2155
Non-Household	
Disconnect/ Non-working	1217
Not a Private Residence	524
Total Non-Household	1741
Unknown	
Don't Know	1
Final No Answer	1081
Language Barrier	130
Final Line Busy	82
Total Unknown	1294
Total Dispositions	5190
CASRO	6.13%
Upper Bound	10.50%
Refusal Rate	28.90%

5. Southeast 2

Household	
Refusal (Eligibility Unknown)	769
Complete	195
No Eligible Respondent at Number	173
Quota Cell Full	16
Final Termination in Questionnaire	18
Total Household	1171
Non-Household	
Not a Private Residence	497
Disconnect/ Non-working	1069
Total Non-Household	1566
Unknown	
Language Barrier	67
Final No Answer	516
Final Line Busy	30
Total Unknown	613
Total Dispositions	3350
CASRO	13.61%
Upper Bound	19.86%
Refusal Rate	23.49%

6. West Coast

Household	
Final Termination in Questionnaire	38
Refusal (Eligibility Unknown)	774
Quota Cell Full	17
Complete	198
No Eligible Respondent at Number	251
Total Household	1278
Non-Household	
Disconnect/ Non-working	815
Not a Private Residence	209
Total Non-Household	1024
Unknown	
Final Line Busy	43
Final No Answer	702
Language Barrier	94
Total Unknown	839
Total Dispositions	3141
CASRO	11.35%
Upper Bound	19.60%
Refusal Rate	25.85%

7. West

Household	
No Eligible Respondent During Interview	38
Final Termination in Questionnaire	21
Refusal (Eligibility Unknown)	399
Complete	199
Quota Cell Full	36
No Eligible Respondent at Number	6
Total Household	699
Non-Household	
Not a Private Residence	28
Disconnect/ Non-working	95
Total Non-Household	123
Unknown	
Don't Know	162
Final No Answer	1137
Language Barrier	35
Final Line Busy	94
Total Unknown	1428
Total Dispositions	2250
CASRO	10.40%
Upper Bound	32.15%
Refusal Rate	18.67%

8. National Sample (U.S.)

Household	
Final Termination in Questionnaire	60
Refusal (Eligibility Unknown)	1675
No Eligible Respondent at Number	116
Quota Cell Full	111
Complete	399
No Eligible Respondent During Interview	5
Total Household	2366
Non-Household	
Not a Private Residence	551
Disconnect/ Non-working	609
Total Non-Household	1160
Unknown	
Language Barrier	212
Final No Answer	1218
Final Line Busy	148
Total Unknown	1578
Total Dispositions	5104
CASRO	11.65%
Upper Bound	18.70%
Refusal Rate	33.99%

APPENDIX C
CHLORINOUS FLAVOR SENSITIVITY TEST INSTRUCTIONS AND ANSWER SHEET

Taste Test Procedure

The objective of these tests is to determine the chlorine concentration that can be tasted and recognized by the population. Chlorine is used in the treatment of drinking water as a disinfectant to kill microorganisms (bacteria and viruses) that may be found in water. Chlorine is used to assure the safety of your water from the drinking water treatment plant, all the way to your home tap. You should not worry about the levels of chlorine you find in your tap water or in your glasses today. These glasses contain safe, potable, drinking water meeting or exceeding all State and Federal regulations. The level of chlorine they contain is within the range of concentrations used in drinking water.

Please, read and carefully follow these instructions:

(1) Three (3) glasses (called "samples") containing drinking water will be given to you.

(2) Starting with Sample A, take a small sip in your mouth. Move sample through your mouth, hold it for several seconds, and discharge it or swallow it. *Please do not remove the foil lids from the samples until you are ready to taste them.*

(3) Repeat Step 2 with Sample B, and then with Sample C.

(4) Select the sample that, according to you, tastes like chlorine (bleach).

Only 1 out of the 3 samples will contain chlorine.

Do not get distracted by a difference in temperature!

The sample that taste chlorine might be difficult to select as some samples have very low concentration. In these cases, record your best guess. You **must** select a sample!

(5) You can taste again any of the 3 samples to confirm your choice.

(6) Record your answer by placing an "X" on your answer sheet in the circle under the sample that taste like chlorine. Do not write in the circles of the samples that do not taste chlorine. This is an example of answer:

	Set #1		**Results**
1A ○	1B ○	1C ⓧ	○

(7) Rinse your mouth with the water labeled "Reference Water" provided in front of you. Rest for 2 minutes.

(8) Another set of 3 samples will be brought to you. Repeat Steps 2 to 7.

You will be tasting 7 sets of 3 samples each.

The Water Utility and Carollo Engineers thank you very much for your time and participation!

Taste Tests

Please, write your name and number after "Panelist".

Location: ______________________________
Date: ______________________________
Group: ______________________________

Panelist: ____

Put an "X" in the circle corresponding to the samples in which you find a chlorine taste.
Please, do not write in the shaded (gray) areas!

Set 1			Results
1A ○	1B ○	1C ○	○

Set 2			Results
2A ○	2B ○	2C ○	○

Set 3			Results
3A ○	3B ○	3C ○	○

Set 4			Results
4A ○	4B ○	4C ○	○

Set 5			Results
5A ○	5B ○	5C ○	○

Set 6			Results
6A ○	6B ○	6C ○	○

Set 7			Results
7A ○	7B ○	7C ○	○

REFERENCES

American Standard for Testing of Materials (ASTM). 1997. *Standard Practice for Determination of Odor and Taste Thresholds by a Forced-Choice Ascending Concentration Series Method of Limits*. E679-91. American National Standard, West Conshohocken, Pennsylvania.

Baker, R.A. 1963. Threshold Odors of Organic Chemicals. *Jour. AWWA*,55(5):913-916.

Baylis, J.R. 1935. *Elimination of Taste and Odor in Water*. MacGraw-Hill Book Company Inc., 392 pp.

Beck, J., J.B. Dyhm, and P. Kieler Jensen. 1986. Preformed Monochlorine Used as a Post-Disinfectant in Drinking Water Treatment at Sjælso Water Works. *Aqua.*

Best Cellars, LLC 1998. Bottled Water Web. [Online]. <http://www.bottledwaterweb.com/indus.html>. [cited 2000]

Bruchet, A. 1999. Solved and Unsolved Cases of Taste and Odor Episodes in the Files of Inspector Cluzeau. *Wat. Sci. Tech.*, 40(6): 15-21.

Bryan, P.E., L.N. Kuzminsky, F. Miles Sawyer, and H.T. Feng. 1973. Taste Thresholds of Halogens In Water. *Jour. AWWA*, 65(5):363-368.

Burlingame, G. 1994. Philadelphia Water Department. Personal Communication.

Burttschell, R.H., A.A. Rosen, F.M. Middleton, and M.B. Ettinger. 1959. Chlorine derivatives of phenol causing taste and odor. *Jour. AWWA*,51(1):205-214.

Cochran, W.Q. 1977. *Sampling Techniques.* 3rd Edition. New York, N.Y.: Wiley.

Cushing, R.S., and G.F. Crozes. 1999. *Guidelines for Evaluating and Controlling Biological Regrowth in the Distribution System*. Denver:Colo.:AwwaRF

De Grunt, F.E., Zoeteman, B.C.J., Piet, G.J. and van der Heuvel, M.P.H. 1977. *Methods for the sensory evaluation of water quality.*

Dietrich, A.M., M.P. Orr, D.L Gallagher, and R.C. Hoehn. 1992. Tastes and Odors Associated with Chlorine Dioxide. *Journal AWWA*, 84(6):82-88.

Doulton Water Filters. 2001. Point-Of-Use Technologies and Applications. www.doulton.ca. INTERNET.

EPA (Environmental Protection Agency). 2001. Factoids: Drinking Water and Ground Water Statistics for 2000. www.epa.gov/safewater; INTERNET.

Faust, S.D., Aly, O.M. 1983. *Chemistry of Water Treatment*. Ann Arbor Science, London.

Gallagher, M.W. 2000. Professional Selling Skills. In *Proceedings of the International Bottled Water Association Annual Convention*. Seattle, Wash.: IBWA.

Gittelman, T.S., and P. Luitweiler. 1996. Tastes and Odors in Treated Water from Algae, Urea, and Industrial Chemicals in Source Waters. In *Proceedings of the AWWA Water Quality Technology Conference.* Denver, Colo.: AWWA.

Henry, G.T. 1990. *Practical Sampling.* Newbury Park, Calif.: Sage.

Hoehn, R.C., A.M. Dietrich, W.S. Farmer, M.P. Orr, R.G. Lee, E.M. Aieta, D.W. Wood III, and G. Gordon. 1990. Household Odors Associated with the Use of Chlorine Dioxide. *Jour. AWWA*.

Hrudey, S.E., S.A Daignault, A.J. Gac, D. Poole, G. Walker, and D.A. Birkholz. 1989. Swampy or Swimming-pool Aldehyde Odors Caused by Chlorination and Chloramination. In *Proceedings of AWWA Seminar on Identification and treatment of Taste and Odor Compounds.* Denver, Colo.: AWWA.

Kajino, M., K. Morizane, T. Umetani, and K. Terashima. 1999. Odors Arising from Ammonia and Amino Acids with Chlorine During Water Treatment. *Wat. Sci. Tech*., 40(6):107-114.

Krasner, S.W., and S.E. Barrett. 1984. Aroma and Flavor Characteristics of Free Chlorine and Chloramines. In *Proceedings of the AWWA Water Quality Technology Conference.* Denver, Colo.: AWWA.

Krasner, S.W. 1988. Flavor-Profile Analysis: An Objective Sensory Technique for the Identification and Treatment of Off-flavors in Drinking Water. *Wat. Sci. Tech*., 20(8/9): 31-36.

Lin, S.D. 1976. Sources of Tastes and Odors in Water. *Water and sewage works*, 123(6):101-104 and (7): 64-67.

Lykins, B.W., J.A. Goodrich, and J.C. Hoff. 1990. Concerns with Using Chlorine-dioxide Disinfection in the USA. *Jour. Aqua*, 39(6): 376-386.

Mackey, E.D., J.E. Davis, L. Boulos, J. C. Brown, and G.F. Crozes. 2003. *Consumer Perceptions of Tap Water, Bottled Water and Filtration Devices.* Denver, Colo.: AwwaRF

Malleret, L. 2001. Developpement d'Outils Analytiques pour l'Analyse de Traces des Composés Responsables des Problèmes de Goût et Odeur de « Terre-moisi » de l'Eau Potable. PhD dissertation, University Paris VI.

Mallevialle, J., and I.H. Suffet. 1987. *Identification and Treatment of Tastes and Odors in Drinking Water*. Denver, Colo:AwwaRF/Lyonnaise des Eaux.

Moncrieff, R.W. 1951. The Chemical Senses. Leonard Hill, London.

Pesticide & Toxic Chemical News. 1998. Organic Food Sales Up Sharply Since 1992, Survey Finds.

Philadelphia Water Department's Bureau of Laboratory Services and the Melior Group, 1986. "1986 Consumer Survey and Taste Test for Philadelphia." Internal report.

Piriou, P., and J. Perrelle. 1999. *Goûts et Odeurs en Réseau d'Eau Potable: Programme 97*, CIRSEE AESN, EP-RSO 9811.

Samples, W.R. 1959. *A Study of the Chlorination of Urea*. Ph.D. diss, Harvard Faculty of Arts and Sciences.

Schlett, C., Pfeifer, B. 1992. Determination of substituted phenols below the threshold value of Odors, *Vom Wasser* (79):65-74.

Schweitzer, L., Suffet, IH., Corado, A. , Ho, J. Antagonistic and synesgistic effects of chlorine on earthy and musty odors in drinking water. Submitted to *Water Science and Technology* 2002.

Suffet, I.H., A. Corado, and D. Khiari. 1993. Brominated Organic Compounds and Taste and Odors in Water Supplies. In *Proceedings of AIDE-IWSA International Workshop, Bromate and Water Treatment.* Paris.

Suffet, I.H., J. Mallevialle, and E. Kawczynski. 1995. *Advances in Taste-and-odor Treatment and Control*. Denver, Colo.: AwwaRF/Lyonnaise des Eaux.

Suffet, I.H., A. Corado, D. Chou, M. McGuire, and S. Butterworth. 1996. Taste and Odor Survey. *Jour. AWWA*:168-181.

Torobin, M.G., T.D. Johnson, and I.H. Suffet. 1999. The Importance of Water Quality Aesthetics in Consumer Confidence in the Safety of Drinking Water Supplies, TU21(5). In *Proceedings of the American Water Works Association Annual Conference.* Chicago, Ill.: AWWA.

Welte, B., and A. Montiel. 1999. Study of the Possible Origins of Chlorinous Taste and Odor Episodes in a Distribution Network. *Wat. Sci. Tech.*, 40(6):257-263.

White, G.C. 1972. *Handbook of Chlorination for Potable Water, Wastewater, Cooling Water, Industrial Processes and Swimming-pool*. Ed. Van Nostrand Reinhold Company.

Whitfield, F.B. 1988. Chemistry of Off-flavors in Marine Organisms. *Wat. Sci. Tech.*, 20(8/9): 63-74.

Wiesenthal K., and I.H.Suffet. 2002. Taste Characteristics of NaCl in The Presence of Free Chlorine or Chloramines In Drinking Water. Submitted To *Water Science and Technology.*

Young, W.F., H. Horth, R. Crane, T. Ogden, and M. Arnott. 1996. Taste and Odor Threshold Concentrations of Potential Potable Water Contaminants. *Water Research*, 30(2): 331-340.

Zhang, G. et al. 1992a. Evolution of Organoleptic Water Quality During Treatment. *Macau. Env. Tecnol.*

Zhang, G. 1992b. *Etude et Maîtrise de la Qualité d'eau de la Ressource à la Distribution à Macao*. Ph.D. diss: Université Paris-Val De Marne.

ABBREVIATIONS AND VARIABLES

ASTM	American Standard for Testing of Materials
AWWA	American Water Works Association
BOM	biodegradable organic matter
BW	bottled water
CASRO	Council of American Survey Research Organizations
CATI	computer-assisted telephone interviewing
CC	combined chlorine
CCR	consumer confidence report
CMOR	Council for Marketing and Opinion Research
CT	concentration x time
DBPs	disinfection by-products
DI	de-ionized
DNA	did not answer
EPA	Environmental Protection Agency
FAQ	frequently asked questions
FC	free chlorine
FPA	flavor profile analysis
FTN	flavor threshold number
FW	filtered water
GW	ground water
HPC	heterotrophic plate count
MCL	maximum contaminant level
MRDL	maximum residual disinfectant level
µg/L	micrograms per liter
mg/L	milligrams per liter
mgd	million gallons per day
MIB	2-methyl isoborneol
MRA	Marketing Research Association
NOM	natural organic matter
NTU	nephalometric turbidity units
P/A	presence/absence
POU/POE	point-of-use/point-of-entry
PR	public relations
PWD	Philadelphia Water Department

QDA	quantitative descriptive analysis
RDD	random digit dialing
SPSS	statistical package for the social sciences
SW	surface water
TCA	thrichloroanisole
TCR	total coliform rule
TDS	total dissolved solids
THMs	trihalomethanes
TOC	total organic carbon
TT	taste test
TW	tap water
TWA	tap water alternatives
U.S.	United States
WTP	Water Treatment Plant

IWA Publishing's authorised EU representative for General Product Safety Regulations is Diane D'Arras, 15 rue Duret, 75116 Paris, France, e-mail: safety@iwap.co.uk.

Printed and bound by CPI Group (UK) Ltd, Croydon, CR0 4YY
17/09/2026
02218799-0001